# THE STORY OF
# POLLINATION

I0090092

B. J. D. MEEUSE
UNIVERSITY OF WASHINGTON

With drawings by Hilda Kern

THE RONALD PRESS COMPANY  ·  NEW YORK

Copyright © 1961 by
THE RONALD PRESS COMPANY

————

*All Rights Reserved*

No part of this book may be reproduced
in any form without permission in writing
from the publisher.

AN ORCHARD INNOVATIONS REPRINT EDITION
Printed in the United States of America

ISBN: 978-1951682330

Ver. 1.0 (4/16/2020)

*From a Dutch uncle, for bringing me sunshine and music and a gentle spring breeze*

Bumblebees on flowers of toothwort, *Lathraea squamaria*.

# *Preface*

An author who has succumbed to the urge to put something on paper has to brace himself for that inevitable question: For whom is your book intended? Let me simply say that this story of pollination has been written for young persons between the ages of eight and eighty (and older). Indeed, if you, the reader, are to find something good in it, you will have to recapture the mood of childhood's constant wondering about this amazing and beautiful world of ours. A sense of poetry will be a great help to you, even though, like a child, you may be unable to define it. I hope that there is some adventure in this book, too, of the type that makes looking through a microscope so attractive and rewarding to some people.

A second inevitable question, I suppose, is: Why did I write it? Well, let's make no bones about the matter—I had a bee in my bonnet. Part of my youth was spent in West Java, a tropical paradise full of birds, flowers, and insects. Even before I could put my feelings into meaningful words, I was fascinated by the contrast between the splendid immobility of the flowers and the darting movements of the butterflies—God's little magicians, as they have been called. Boldness and intimacy, brilliant colors and bizarre shapes! Gradually I began to realize, dimly at first, that observation of these beautiful creatures would yield that almost unique reward: a blend of intellectual and poetical satisfaction. Strange to say, however, it was impossible to find any real guidance when I tried to become familiar with the creatures around me. My guess is that the scientists who at that time were so abundant in my home town were perfectionists, unable to produce the unpretentious books that would have been the only real help to green boys like myself. In passing, I venture to point out that most of them never did produce the perfect books they must have had in mind. "Better" things are always the enemy of good things.

This childhood experience has goaded me to share with others what knowledge I now possess about the wonders of pollination. May I be forgiven if a missionary's zeal has sometimes carried me away. Although some readers may think that I have occasionally gone off on tangents, I hasten to assure them that there is some system in my madness. In the first part of the book I have tried to present, in a more or less logical order, a number of principles— or at least matters of general importance—concerning the relationships between flowers and the animals that pollinate them. Thus, there is a chapter on the color vision of animals, one on the ways in which flowers produce their color effects, and a chapter on the honey-guides that lead the pollinators to the hidden nectar.

Even a brief inspection of these earlier chapters will show that it was impossible for me to avoid the introduction of special animals such as the beefly which, unknowingly, have played such an important role in the establishment of our general principles of pollination. I do not apologize for this because, after all, what is a story without heroes? Unashamedly, I have included in the latter half of the book chapters on the most important pollinators—bees, birds, butterflies, hawkmoths, flies, bats, and beetles. Some of the matters covered in these chapters have, as I say, already been taken up earlier in the book, but the story is now told more or less from the viewpoint of the pollinating animal, a change in emphasis which has its advantages, too.

In general, I have tried to let the facts do the talking. This means that there will be enthusiasm here, but no sticky sentimentality, for it is always wrong to read human motives into the actions of animals. In those few cases where the reader may get the impression that I have "humanized" plants and animals a little, he should remember that it was done with tongue in cheek.

I do not believe that this is a book which should be read in one sitting. Rather, in order to give the words real life, the reader should try to intersperse trips into the field for observation. Such trips, undoubtedly, will raise many questions in his mind. A majority of these are answerable, but the forgiving reader should not expect to find all the answers here. For reasons which by this time must be clear, I have made no attempt to write a book that contains all the answers.

It is a privilege to thank all those friends who, in some way or other, have encouraged me in the writing of this book. Particularly do I thank Niko Tinbergen of Oxford University, a guide in some of my earlier steps as a scientist. I am deeply grateful to my wife who, without complaint, saw me devote to this book many hours which might have been spent more companionably. My warm appreciation goes to Miss Hilda Kern for her skillful and patient execution of the drawings, and to Ben Goffe for the comradeship and splendid help he gave me in taking the photographs.

B. J. D. MEEUSE

Seattle
August, 1961

# Contents

*Part I*

# OF RHYME
# AND REASON

# 1

## *Dawn*

Do you know Emily Dickinson's little poem "Prairie"? The first lines go like this:

> To make a prairie it takes a clover and one bee,—
> One clover, and a bee,
> And reverie.

I love it; therefore, you can be sure that I am not trying to be funny when I turn things around a little bit to point out that "To make a reverie, it takes one clover and a bee." Yes—these two can give one plenty to think about! Let us just go down to a field of red clover on a nice bright day in the summer when the plants are in full bloom. From a distance you can already guess that it must be quite a lively place. Butterflies, such as cabbagewhites and the beautiful yellow and orange "sulphurs" are tumbling up and down over the field in their gay little fashion. However, it is only when we are quite close that we can truly appreciate the liveliness of the place: it is simply abuzz with bees.

The bumblebees are especially well represented. These animals are bigger, fuzzier, and much more colorful than their cousins, the honeybees in their modest orange-brown attire. Striking color combinations of black with yellow and white or orange seem to be favored by them. Since neither they nor the honeybees are very shy, we have no trouble in watching them as they visit one flowerhead after another, skipping hardly any of those little florets in each head which are open. Knowing that one acre of red clover brings forth more than 200 million of these florets, one can see that the bumblebees must really work hard to get them all.

I hope you are interested enough to follow an individual bee with your eyes for some time. You will then find that the animal can take care of 300 to 400 florets in half an hour. Let me ask you to pay special attention to its legs during these wanderings. It is not hard to see that a bee has six legs (three on each side) and that the last pair is broader and flatter than the others. But these broad legs seem to keep changing all the time, at least in some individuals: clumps of a yellow material will appear here, growing thicker after each visit to a flower, until finally they bulge out like full saddlebags.

Nowadays, even a city child knows what these clumps are, and what the bees are doing on the clover (or on other flowers, for that matter). Keen on the food which the flowers provide, the insects are inadvertently doing them a good turn by taking care of the process of pollination. Just to refresh our memory, I have pictured the chain of events in a diagram (Fig. 1). It can be seen that in a typical flower we find a pistil surrounded by stamens. The thick upper part of each stamen, the anther, produces a huge number of tiny grains, forming a fine powder which we call pollen. Most often, the powder is yellow or orange, but occasionally we encounter a flower with blue or almost black pollen. Each species of flower has its own characteristic pollen color. In the bottom part of the pistil we find the ovules, which we might call the beginnings of the seeds. For these ovules to develop and really form seeds, however, it is necessary that, in some way or other, pollen grains get to the sticky upper end, the stigma, of the pistil. Sometimes the wind will do the job, in other cases insects, birds, bats, man—or even water. It is on the stigma that the pollen grains will germinate, which means that a tube will begin to grow downward from each one, through the thin part of the pistil which is called the style. When such a tube finally reaches an ovule, certain of its "ingredients" will unite with certain "ingredients" of the ovule. In botany books, one can find this rather complicated event described under the name of "double fertilization." But let us not worry about details. We simply notice that it is just as if the ovule has been given the green light, and that it will now rapidly transform itself into a seed.

Yes, nowadays many city children know this, and usually a great deal more, too. For instance, they know that in many cases there is no cooperation between pollen grains and pistil in one and the

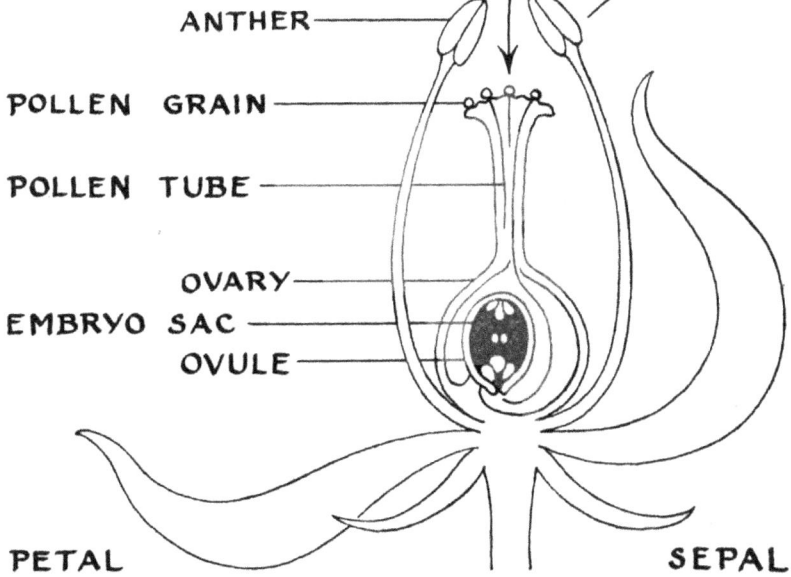

SELF-
POLLINATION

CROSS-POLLINATION
BY WIND

BY ANIMALS

ANTHER

POLLEN GRAIN

POLLEN TUBE

OVARY

EMBRYO SAC

OVULE

PETAL

SEPAL

Figure 1. Diagram of self- and cross-pollination.

5

same flower. As the great Charles Darwin (1809–1882) put it, "nature seems to abhor self-pollination," and in order to get good, viable seeds it is much better when the grains are transported to a flower on another plant of the same species. This is what we mean when we talk about "cross-pollination." Fig. 2 shows the trick

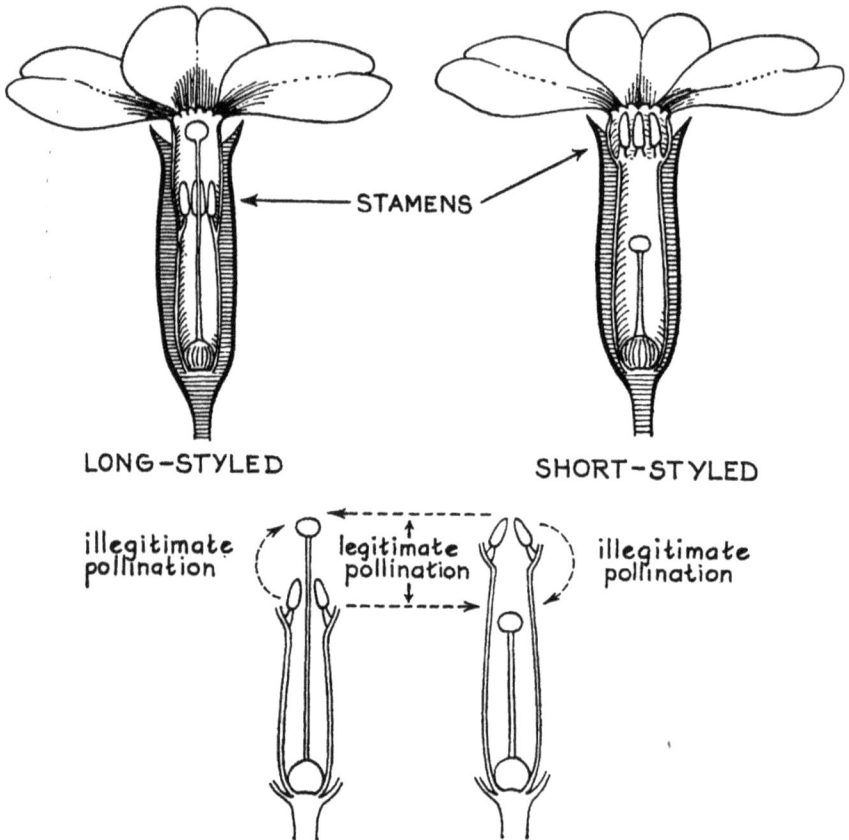

LONG–STYLED                              SHORT–STYLED

illegitimate           legitimate            illegitimate
pollination            pollination           pollination

Figure 2. Long- and short-styled form of the common garden primrose (*Primula*), with pollination scheme. Self-pollination is made very difficult or impossible.

which primroses employ to make sure that it takes place. They produce two types of flowers, one with short stamens and a long pistil, and another one where it is just the other way around. One can immediately see that insects are absolutely essential here. But our red clover is an excellent case in point, too. It also happens to be one of those plants which, for good seed production, depend upon cross-pollination, and that process, in its turn, is dependent on the presence of enough bumblebees.

The clumps of yellow material which we noticed on these animals were collections of pollen grains; the flat or slightly hollow parts on the bee's hind legs, which so obviously serve the purpose of carrying the grains, are called pollen-baskets. In a later chapter we shall see that, for their young, bumblebees need the sweet nectar which many flowers produce, as well as pollen. The fact that in the case of clover this precious fluid is hidden rather deeply at the base of the flower does not cause any special trouble to a bumblebee, since it is the fortunate owner of a long proboscis or tongue. In the act of securing both nectar and pollen, the animal is perfectly able to do a marvelous job of cross-pollination. Honeybees, as we have seen, are common visitors to clover fields too, but their tongues are much shorter and therefore they just do not compare with bumblebees in efficiency, in this particular case. The difference was beautifully illustrated in the nineteenth century when red clover was introduced into New Zealand, a country where bumblebees originally were lacking. No good seed crops could be obtained until finally, around 1880, about 100 specimens of a British bumblebee, *Bombus terrestris*, were introduced, too. The species soon became established, and the results on clover were quite dramatic.

The importance of bumblebees has also been illustrated facetiously—by Darwin, I believe—by pointing out that the greatness of the British Empire depended on its spinsters. The argument goes approximately as follows: spinsters keep cats; cats keep down the population of mice that are so destructive to bumblebee nests; consequently, the bumblebees flourish, and there will be an abundance of clover to feed the cattle; this, finally, ensures a good supply of the red meat eaten by Her Majesty's sailors and soldiers.

But all joking aside, have you ever stopped to think what a dull and different world we would have if our pollinating insects dropped out of the picture? Surely, we would still have the fruits of grasses such as wheat and corn, plants that are either cross-pollinated by the wind or undergo self-pollination. Likewise, most of our legumes such as peas and beans would still be available, since their usual mode of pollination is selfing in unopened, young flowers. On the other hand it would be practically impossible to grow alfalfa and zucchini, and the yields of many other crops, such as buckwheat, cotton, and onions, would be terribly reduced. In all seasons of the year, the sad effects would make themselves felt. Gone would be the gay carpets of spring flowers. In the summer, we would have to do without sweet cherries, watermelons, and cucumbers. Thanksgiving would be a dismal affair without pumpkin pie, almonds, and apple cider, and even Christmas would be changed, since holly trees will only produce good, commercial red berries after cross-pollination. Many soil-holding plants, too, would simply be wiped out, so that erosion would become an even more terrible problem than it already is.

In view of all the things I have just said, it is downright funny to realize that only two or three centuries ago people knew next to nothing about pollination. Certainly, there were many who loved flowers, grew them in greenhouses, made beautiful paintings of them, and even risked their lives (sometimes) for getting new or rare kinds, but they were all blissfully ignorant of the role of pollen grains and ovules. Of course the beekeepers could not fail to notice the clumps of pollen grains which their bees brought home from the field in their pollen baskets, but they did not know what to make of them. Some thought they were tiny pebbles, picked up deliberately by the animals to give them better equilibrium in flight.

Yet, a long time ago, there were at least some people with a "working knowledge" of pollination. In the Near East, dates have always been an important human food, and the wealth of a man was often expressed in the number of date palms he possessed. There are two types of these palms: the males, producing flowers with only stamens and pollen, and the females which have flowers with only pistils. Pollination, which is taken care of by the wind, presents no problems as long as there are enough female and male trees around. If, however, the latter are too heavily outnumbered, it is wise to cut off the few male inflorescences and shake them back and forth in the immediate vicinity of the female trees so that the pollen may be scattered. Figure 3, copied from an ancient Mesopotamian relief, shows that in the days of King Ashurnasirpal (884–860 B.C.) people must have been familiar with this practice. The Greeks also knew the story in a fairly accurate way.

Contrary to what one would expect, however, the idea that there is something male and something female in plants as well as in animals was soon forgotten completely—at least in the Western world. The whole affair had to be rediscovered, and of course a more scientific basis had to be given to it. Credit for this should go to a number of people, including the Germans Rudolf Jakob Camerarius (1665–1721) and Joseph Gottlieb Koelreuter (1733–1806), the Irishman Arthur Dobbs (around 1750), and the great Swede Carolus Linnaeus (1707–1778). Koelreuter, especially, made a magnificent contribution, but he may have been a little over-impressed by the idea that the pollen will get to the pistil without

Figure 3. Mesopotamian relief dating back to the days of King Ashurnasirpal (884–860 B.C.). Two deities, holding male inflorescences of the date palm, are engaged in the artificial pollination of a female date tree, represented in highly stylized form in the center. (After an original in the Leiden Museum of Antiquities.)

any particular trouble—by self-pollination or by the wind. It is certainly true that there are some plants, such as shinweed (*Pirola*) where this is the case; self-pollination here is the most natural thing in the world because of the peculiar position of stamens and pistil. As to the wind, we have already seen what it will do in the case of palms. And who has never observed the clouds of pollen over an evergreen forest in the spring, or over a wheat field in June?

Still, there are countless plants where the wind cannot do very much good because the stamens do not stick out far enough and the pollen is sticky. It is here that insects or other animals must come to the rescue. Although Koelreuter was not blind to the role of insects, it is fair to say that Christian Conrad Sprengel (1750–1816) was the first to really see the light here; therefore, he will, in a way, be the hero of my story.

# 2

# *Sprengel's Superb Divinations*

When we used the rather Biblical expression "to see the light" in connection with Sprengel, we did not do this facetiously. He was a deeply religious man who saw the hand of the Creator in the minutest details which he observed in flowers. He was convinced that even the smallest color spot, the tiniest little hair was there for a special purpose. To himself, his wonderful discoveries were something very much in the nature of a revelation, and the delightful book in which he describes his "flower adventures" is aglow with this spirit, humble and joyous at the same time. Let us follow the account which he himself has given in that book, *Das neu entdeckte Geheimniss der Natur*—the secret of nature revealed:

In the year 1787, Sprengel examined very carefully the flowers of a wild geranium, *Geranium sylvaticum*. At the base of each petal, on the hollow inside and also along the margin, he noticed a number of soft, small hairs and, true to his character, began to wonder why the Creator had placed them there. And then, in a sudden flash of insight, he realized that there must be a connection with the drops of sweet fluid which were also present, one on each petal, secreted by a special gland. Could it be that these drops of "nectar" were there to attract insects? If that was the case, then it would only be natural that they had to be protected against the ill effects of rain water, which would dilute them. The hairs were there, then, for that special purpose—to protect the nectar drops against rain.

The next year, he examined the flowers of a forget-me-not, *Myosotis palustris*. Sure enough, here also he found a sweet sap

11

(although it was more deeply hidden) and a number of hairs for
its protection. He also noticed the yellow "ring" around the entrance to the flower tube, standing out in such a pretty way against
the sky blue of the petals (Fig. 4). Could it be, he thought, that
this patch of different color was there to serve the insects as a
guide, to point the way (so to speak) to the hidden nectar? He
examined a large number of different flowers and found that his
idea was sound. In a great many cases, there were patches of a
different shade, and sometimes patterns of dots or lines which really
indicated the hidden treasure. Sprengel invented the word "Saft-
male," that is, "sap signs" or "sap marks," for such spots. In
English-speaking countries, we usually call them "honey-guides" or
"nectar-guides." With impeccable logic, he then made the next step
in his reasoning. If these honey-guides were indeed there for the
benefit of insects, then the color of flowers, as a whole, must also
be there for their sake; the purpose of the color (or, as we would
nowadays say, their function) must be to attract insects to the
flowers from a distance.

The next year Sprengel studied irises and found that the flowers
of these plants, for their pollination, have to rely almost entirely
on the activities of bees and bumblebees (Fig. 5). Better still, he
discovered that there are many cases in which a plant species is
pollinated by one single species of insect only. Still a year later, he
discovered the fact that in the flowers of fireweed, *Epilobium*, the
stamens reach maturity before the pistil does, so that self-pollination cannot easily occur (it does take place eventually when cross-
pollination is not achieved, but that is another story). Nowadays,
we would say that the flowers of fireweed are proteranderous.

---

Figure 4. Flowers with and without nectar-guides. Left-hand column, upward: a monkey-flower (*Mimulus*), showing a system of dots; a hawkmoth-pollinated honeysuckle (*Lonicera*), lacking nectar-guides entirely; a garden primrose
(*Primula*), with yellow patch on purple background; a rhododendron (*Rhododendron*), showing dark color patch and dots. Right-hand column, upward: evening-
primrose (*Oenothera*), hawkmoth-pollinated and devoid of honey-guides; Virginia bluebell (*Mertensia*), yellow lines in middle of petals (1½×); foxglove
(*Digitalis*), system of dots; forget-me-not (*Myosotis*), yellow ring around flower
entrance (6×). Middle column, upward: pansy (*Viola*), showing a combination
of color patch and radiating lines; toadflax or butter-and-eggs (*Linaria*), orange
patch against light-yellow background; bindweed (*Convolvulus arvensis*), spokes
of a wheel; Kenilworth ivy (*Linaria Cymbalaria*), yellow spot on purple flower
(4×).

Figure 4

Later, Sprengel found just the opposite situation, proterogyny, in a *Euphorbia* species where the pistil reaches maturity earlier than the stamens. The result is of course the same: self-pollination cannot take place.

It is mainly on these observations that Sprengel founded his beautiful cross-pollination theory. Yet he did not immediately get the credit he so richly deserves. In part, this may have been his own fault. So full was he of the beauty of his revelations that he was like a man possessed. He thought that if he shared his new-found riches with other people it would make them just as happy as he had been, and how disappointed he was when it turned out that this was not so! Of course, one can hardly blame people for not showing too much interest. It was, after all, a time of bloodshed and upheaval; the French Revolution had just begun. But Sprengel could not understand how people could do such trivial things as eating, gossiping, and sleeping when such beautiful events were taking place around them in nature all the time. He began to neglect his job of Rector at the dignified Spandau school which was known under the name of "Fredericianum"; sometimes he would even skip the Sunday services. Of course, such an irreligious thing could not be tolerated! So, this man had to quit, and from then on he had to make a living by giving private language lessons and organizing lectures or field trips. One could participate in the latter by paying 2 to 3 groschen per hour—the equivalent, probably, of 2 or 3 pennies. I know that money was worth more in those days than it is now; still, his income did not exactly enable Sprengel "to live it up" and he died almost penniless. I do not believe that we possess a single portrait of this man. Our only comforting thought in this distressing affair is that to his dying day he must have had a song in his heart.

Darwin, who independently rediscovered many of the things which Sprengel had noticed, and who heard about his book afterwards, has shown his stature by drawing attention to Sprengel's work, giving him the prominent position in science which he still occupies today. Also (and I am sure that Sprengel, had he been alive, would have liked this even better), Darwin continued to build on the foundation that had been laid; his book on the pollination of orchids (1862) is a jewel. And others have come after him: men such as Federico Delpino in Italy, Anton Kerner in

Austria, Fritz Müller in Brazil, Ernst Loew, Paul Knuth, and Hermann Müller in Germany, Asa Gray, Charles Robertson, and William Trelease in the United States—all inspired by the great examples that had been set. In this little book, we shall have plenty of opportunity to deal with their work when we discuss the varied and wonderful contraptions of certain flowers.

After what I have said, I do not think that there is any doubt about our feelings for Sprengel. And yet, we have to concede that he did not really prove his point. Sprengel had that poetic and intuitive insight which we can rightly call genius; his work is like the lightning thrust of a rapier, penetrating right to the heart of the matter. But the weak point is that he ascribed to his insects powers of discrimination which they might not always possess. Is it true that the colors of flowers mean something to them? Do they distinguish between different colors and are they attracted by some? Do they allow themselves to be led by Sprengel's nectar-guides? What does the scent of flowers mean to them, and do they really taste nectar as something "sweet"? Let us consider these, and other, questions in the light of modern experiments and developments. In doing that, we shall try to take up, one by one, the various points raised by Sprengel. Color vision, then, is the first topic we have to discuss.

Figure 5. Bumblebee pollinating a flower of *Iris sibirica*.

Figure 5

# 3

# A Rainbow of Colors

It was a sad day for the many people who had begun to believe in Sprengel's theories when Carl von Hess, in 1912, published the results of his experiments on the color vision of bees. His conclusion was that they are utterly and completely color-blind. This would of course mean that all Sprengel's wonderful speculations about the pretty colors of flowers and about the honey-guides were utter nonsense. And the worst part of it was that it seemed impossible for an honest person to reject the cold, hard logic of von Hess's experiments—they were too good, or, at least, so it seemed on the surface. This business is important enough for us to pause for a moment and examine von Hess's methods in some detail.

Of course it is always a difficult thing (to say the least) to find out what impression a certain color makes on a creature different from yourself, even though this creature may be your own brother. How can we ever be sure that he will undergo the sensation "yellow" or "blue" in just the same way as you do? Well, at least we can talk to him about these matters, but in the case of animals, that is impossible and we are restricted to roundabout methods. Now, it so happened that von Hess, who was an ophthalmologist, had considerable experience with people that were completely color-blind. He knew that such persons, looking at the band of rainbow colors produced by the passage of white light through a glass prism, will claim that the region from yellow-green to green is "brightest." (Brightness, by the way, is the only thing that counts in the case of totally color-blind persons; as far as they are concerned, the world might just as well show various shades of gray only.) To a normal person, on the other hand, yellow will seem brightest. Von

Hess also knew that bees confined in a narrow space such as a box will move toward the spot that seems brightest to them. Lo and behold, when given the choice, they would move to the yellow-green and not to the yellow region of the "spectrum." This is exactly what a color-blind person with the mentality of a bee would have done. There was also a similarity between bees and totally color-blind persons in their behavior toward certain red colors, namely the ones very far over to one side of the prism-produced "spectrum." They could not distinguish these red colors from black, which is just another way of saying that they were red-blind.

Von Hess summarized his results in a beautiful, sweeping statement to the effect that we could now once and for all exclude the possibility that bees have a color sense comparable with ours. It is easy to imagine the general feeling of depression which his words caused in the ranks of insect and flower lovers. Fortunately for Sprengel, there was one important thing which von Hess had overlooked: *an insect may not always be in the mood to pay much attention to colors—even though it may perfectly well be able to distinguish between them.*

I do not think there is a better example for demonstrating this than the behavior of one of my favorite butterflies, the grayling. This creature belongs to the family of the meadowbrowns and is quite common in western Europe. The way in which Nature has camouflaged this animal is something fantastic; finding a grayling that is resting on a sandy path amidst mosses and lichens is just as impossible as discovering it on the bark of a birch tree. However, in the late morning and early afternoon hours of bright, warm days, the males will invariably give themselves away by their habit of darting after moving objects and following them. A falling leaf, another butterfly that passes, a dragonfly—all these things (and even their shadows) attract their attention. About 20 years ago, we discovered that a male, in doing this, is not just being playful or pugnacious; no, it is his mating urge that goads him on. Naturally, if he keeps darting after moving objects often enough, he may eventually chance upon a female of his own species, and this then may lead to interesting developments which do not need further explanation here.

The question that interested us was: how closely should the moving object resemble a female grayling? In other words, how different can it be before the male loses his interest? So we proceeded to produce paper models which we would tie to a piece of string and dangle in front of the male to see if this would provoke a response. We tried models that had the shape of a butterfly, as well as circles, squares, and rectangles; huge and tiny models, models in different colors, gray, white, and black ones. We varied the movement of the models—in short, we gave all the possibilities that occurred to us a try. One of the most striking results of these experiments was that the color did not really matter. To be sure, we got the impression that a dark object was favored over a light one, but there were no differences, for instance, between blue, green, yellow, and brown. So, for all practical purposes the male graylings behaved as if they were completely color-blind. And yet these same butterflies, enclosed in a big, bright cage, would spontaneously visit colored papers lying on the bottom of that cage, and would always show a very marked preference for the blues and the yellows (Fig. 6). They would distinguish them from gray papers with the greatest of ease. In short, they behaved in the manner which one would expect from an animal that in nature gets the greater part of its food from colored flowers. The moral of the whole story is that an animal which *behaves* as if it were color-blind is not necessarily color-blind. Also, when we study the reactions of an animal, we should never rely on one experimental setup only.

I hasten to point out that a long time before we performed our experiments with the grayling, Dora Ilse in Germany had already done decisive work on the color vision of butterflies. She, in her turn, had only followed the example of that great Austrian pioneer, Karl von Frisch—the scientist who was the first one to show von Hess wrong. Let us briefly examine the ingenious methods he followed when he studied the color vision of bees.

Some time before von Frisch started his experiments on bees, the psychologist Hering had already done good work on the color sense in humans. On the basis of that work, a series of papers in standard colors had been produced, the famous Hering series, in which No. 1 is a red color, No. 4 a golden yellow, No. 12 a certain shade of blue, etc. In addition, there was a Hering series of some 30 gray papers, ranging all the way from white to the deepest black. All these papers were (and are) commercially available.

Von Frisch soon found that it is possible to accustom, or train, bees to a certain color, let us say Hering No. 4, by putting a piece of paper in that color on an experimental table some 100 yards from the beehive and "scenting" it with a drop of natural honey. Very soon, one or two bees attracted by the smell will sip up some honey and return to the hive, where they give it to other bees, with the result that before long scores of bees will visit the yellow paper again and again. This was allowed to continue for quite a while, but then the yellow paper was replaced by a whole group of square pieces of Hering paper, no less than 30 of them in all the various shades of gray, arranged in a checkerboard pattern, with 2 squares of yellow paper placed among them in random fashion. In order to eliminate possible effects of smell, a glass plate was laid on the papers. This had the added advantage that possible differences in glossiness were eliminated. On each square, von Frisch placed a watch glass with some sugar water (which is without smell). The result of this experiment was gratifying indeed: the bees, returning to the experimental table, would almost invariably pick out the yellow papers. That the exact *place* of the yellows in the checkerboard was not important could be shown easily enough by moving them to another position; the bees would follow suit and would still visit them in preference to the grays. As an extra check, experiments were then performed in which the watch glasses on the yellow papers were left empty, whereas those on the grays still contained sugar water. Still, the bees would start looking for a food source only on the yellow papers.

The bees, then, can distinguish this particular yellow (No. 4) from all shades of gray. In other words, they see it *as a color.* In other experiments that were done in very much the same way, von Frisch could show that Hering No. 13 (a blue) is seen as a color. But Hering No. 1, a scarlet red, was confused with black and various very dark grays, showing that bees are red-blind. Likewise, green-blue (Hering No. 10 and No. 11) was confused with shades of gray of medium darkness. So, this color—which in nature is very seldom found in flowers—cannot be accepted as such by the bees.

The next question is: how delicate is the bees' color sense? Can they tell apart two colors which to our eyes seem to be very close and yet different? It is well known that the human eye can distinguish some 60 different colors. To answer our question, we train

Figure 6. Color vision in the grayling, one of the meadowbrown butterflies. The animal pays spontaneous visits to colored papers, favoring blues and yellows while disregarding pure reds and greens.

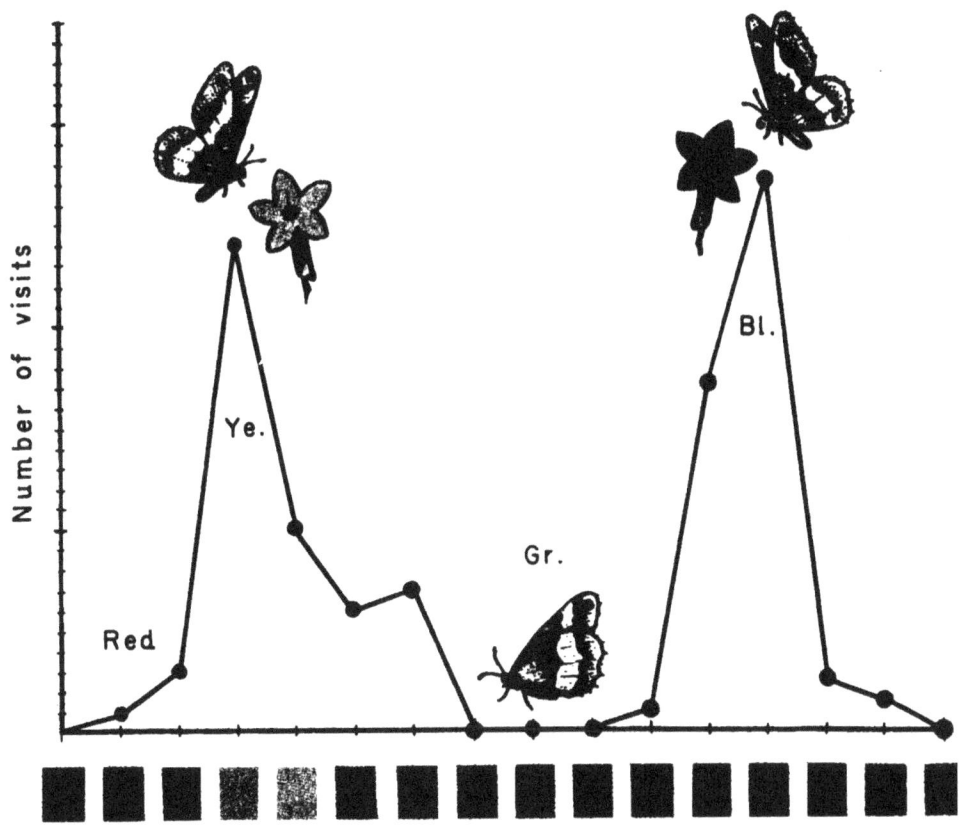

Figure 6

the bees to a special color, say blue No. 13, and then offer them a choice between this particular blue and all the other colored cards which we happen to have, arranged in the now familiar checkerboard pattern. Now it turns out that the bees are not so sure. They visit No. 13 all right but in addition pay calls to the violet and the purple squares. If we train bees on yellow, they will in the crucial experiment also visit orange and green papers. Stranger still, if we train them on orange they will alight more often on yellow than they do on orange; if we accustom them to green, they still prefer yellow. From these experiments we must conclude that (to bees) orange, yellow, and green represent one single color, and that the same is true for blue, purple, and violet. The reason why they prefer yellow to the other colors in its "group" must be that our yellow papers to them must seem more "saturated." This means that, to them, it is a "purer" color than the others, with less admixture of white. Likewise, it can be shown that in the blue group the bees discriminate somewhat against the violet cards, preferring the blue and purple ones.

Von Frisch's experiments were repeated, and beautifully confirmed, by Professor A. Kühn and his collaborator F. Pohl. But instead of using squares of colored papers, Kühn decided to train his bees on the colors of a spectrum, which is a definite improvement because these are always better defined, and purer, than the colors of standard papers. It now turned out that there is a very special region in the blue-green part of the spectrum which the bees recognize as a color, distinct from both "yellow" and "blue." Also, and this is more amazing, the bees are able to recognize the ultraviolet as a color, which is more than a human can do. But then it is to be remembered that man can beat them in the red part of the spectrum. So, when we compare the parts of the spectrum that are visible to a bee's eye and to ours (Fig. 7), it is just as if there has been a shift toward the ultraviolet in the case of the bees. Roughly speaking, the human eye is sensitive to wave lengths ranging from 800 millimicra (red) to 400 millimicra (violet), the bee's eye to wave lengths between 300 millimicra (ultraviolet) and 650 millimicra (orange). One millimicron represents 0.0000001 mm or 1/25,000,000 inch.

So, the upshot of the experiments of von Frisch and Kühn is that, in its visible "range," the bee's eye distinguishes only four different color qualities. Mathilde Hertz, who in 1938 returned to this topic of color vision in bees, suggested calling these colors: bee's yellow (650–500 millimicra), bee's green (500–480 millimicra), bee's blue (480–400 millimicra), and bee's red (400–300 millimicra). If we assume that these results are valid for other insects, too, we can say that they put us in a position to explain certain well-known facts very nicely. For instance, if all insects are red-blind like bees, we can understand at once why flowers of a pure red are so rare in Europe. After all, insects are practically the only pollinating animals there. In America, Africa, and Asia, on the other hand, we find

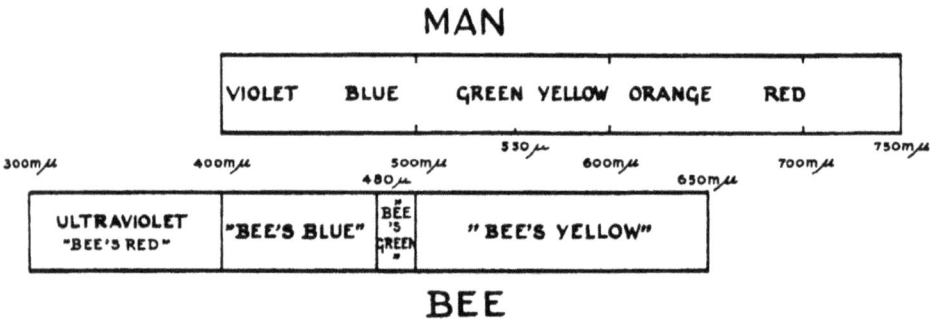

Figure 7. Comparison of the color vision of man and honeybee.

many scarlet-red flowers. A good example is *Salvia splendens,* from Brazil, which has found its way to the parks and gardens of more temperate regions. In the continents just mentioned, and especially in America, birds play an important role in pollination. And we know that there is not too much difference between the color vision of man and that of birds, with one possible exception: the sensitive layer or retina in the eyes of birds always contains a number of fat globules which are yellow to red in color. We cannot, hence, escape the idea that together they act as a "red filter"; that is, it is just as if the bird were wearing red spectacles. Therefore, it is not at all amazing that the birds whose color vision has thus far been investigated are not quite as sensitive to blue and violet as are humans, whereas they react much more strongly to red colors than we do. To be sure, it has to be admitted that we lack sufficient experimental evidence concerning flower-pollinating birds, but perhaps it is not without significance that people who feed hummingbirds sugar water from glass tubes, as a hobby, attract them by attaching red ribbons to these tubes.

Red poppies, at first sight, may cause some puzzlement. They are definitely European, and yet they are scarlet red. Also, they are quite common; we only have to remember the poppies on the battlefields of the First World War, so numerous and well known that they are now used as a symbol on Poppy Day. Bees are very partial to poppies, but it can be shown that they are attracted by the *ultraviolet* light which the poppies reflect and not by the red. Both man and bee, then, see poppies as colored flowers, but in a different way.

The other "red" flowers which we find in Europe are mostly purple rather than red, and they appear blue to bees. The few exceptions to this rule, such as certain European pinks and silenes, are not pollinated by bees but by butterflies. Soon we shall see that at least some of these animals have the ability to distinguish red as a color.

Some inconspicuous, odorless, greenish-white to green flowers such as those of bilberry and Virginia creeper also get numerous visits from bees, a fact that seems incomprehensible after what we have reported earlier. But then, we have good reason to suspect that these flowers reflect the ultraviolet rays and are far from inconspicuous to bees. Speaking of this ultraviolet reflection, we have to devote special attention to white flowers. What we usually call "white light" is in reality a certain combination of the various colored rays which we find in a rainbow. This we can show very easily by passing "white" light first through a prism, so that we get the well-known spectrum of rainbow colors, and then through a suitable lens which reunites these colors and gives us white once more. What would happen if we kept back one of the rainbow colors, say blue, before we reunited the rays with the aid of the lens? The resulting light would seem *yellow* to us. We sum up this situation by saying that blue and yellow are complementary colors. In the same way, it can be shown that red and green are complementary; remove the red rays from the spectrum, and green results, or vice versa. In general, the removal of any color from white light should result in "colored" light.

Although we really do not know very much about the way an insect's eyes operate, there is at least a chance that we can apply the same idea to them. A combination of all the wave lengths which an insect such as a bee can see must, then, give it the impression of "white." Removal of the color "ultraviolet" will leave a light that is no longer white but has a color complementary to ultraviolet, probably blue-green. It is an interesting fact that the natural flowers which we call "white" usually absorb (that is, destroy) ultraviolet rays. So, these latter rays are lacking in the light reflected from these flowers; bees get the impression of something colored and will be attracted.

Perhaps we can generalize this situation, too, for in his experiments on the fly, *Bombylius*, and the hawkmoth, *Macroglossa*, Fritz Knoll has found that white flowers were sometimes confused with flowers stained with methylviolet. A flower that would reflect *all* the wave lengths visible to bees and would really be white to them, would not interest them very much. This was shown in a beautiful way by Mathilde Hertz in experiments with various types of "white" paper. "White," here, means that these papers were white to human vision. Bees could sometimes be trained to these papers with great ease and sometimes only with the greatest difficulty or not at all. Invariably, the papers to which bees could be trained turned out to be those which absorbed ultraviolet rays and appeared "colored" to the animals. The other papers reflected all the rays, including the ultraviolet, and were therefore really of a "bee's white." The insects could never learn to select these with certainty from a checkerboard of papers in various shades of gray.

The beauty of von Frisch's work was soon recognized generally, but at the same time it was realized that his research had to be duplicated with other insect forms before generalizations could be made with safety. This is why scientists such as Fritz Knoll, Dora Ilse, and Hans Kugler started working with such animals as flies, hawkmoths, butterflies, bumblebees, and yellowjackets. It is a pleasure to report on their work, because this gives me an opportunity to introduce several interesting insects which readers with a serious interest in flowers and gardens would probably like to meet anyway.

In 1921, Knoll began a series of studies which he called "Insects and Flowers." The first insect he chose to work with was a furry-looking fly, *Bombylius fuliginosus*, about one-third of an inch long, with a slender proboscis. In German-speaking countries people call it the "Wollschweber," or woolly hoverfly. For the sake of brevity, we shall refer to it simply as *Bombylius* or beefly.

Closely related forms, some with a proboscis at least half an inch long, are common in America. There is one which I find regularly visiting dandelions on my lawn in Seattle in the springtime, and another one that can be seen in June, frequenting monkey-flowers and erigerons along the rocky coast of the Point Lobos peninsula in California. It is always a delight to see these woolly fellows at work, for they show a dash and elegance that is not too common in flies. Although their coat of hair makes them look like bumble-bees, their behavior is very much like that of a hummingbird—at least they have the same habit of sometimes remaining motionless in the air.

The flowers for which *Bombylius* has a preference usually have strongly pronounced colors; for instance, yellow, blue, or violet. Knoll did most of his *Bombylius* experiments in Central Europe with the aid of the attractive blue grape-hyacinth, *Muscari comosum,* which grows wild there. This plant, by the way, is well known in America, too, as an escape. I know of at least one place in Pennsylvania where the meadows in the spring are blue with grape-hyacinths.

The inflorescence of the grape-hyacinth emits a wonderful odor, and one of the first problems which Knoll tried to solve was whether the color or the smell attracts the hungry *Bombylius* from a distance. Just by observing the animals for a while, one can become convinced that it is the color, for they approach the flowers in a straight line like an arrow, without paying much attention to the wind. However, we want to prove our point to the hilt, and therefore it is better to follow Knoll's procedure faithfully. First of all, then, it is necessary to determine accurately the direction of the wind. This we can do by setting up among the flowering grape-hyacinths small gallows, each with a little piece of fluffy, light material attached to it with a piece of thread. Usually, it is not at all difficult to find that light material; we just take the hairtufts from dandelion seeds or from a coltsfoot's "candle" growing nearby. In this way, we soon find the direction and width of the "odor stream" emanating from the inflorescence. Furthermore, we can to a certain extent separate the "optical signals" (the colors), which the flowers use for advertisement, from the "smell signals." We simply invert a wide glass tube over the inflorescence so that the tube rests on a stick and the odor seems to come from the base of the plant (Fig. 8). Finally, we can produce "optical illusions" by introducing into the glass tube colored papers such as Hering No. 4, a golden yellow, which conceals the blue inflorescence.

The result of all this is very striking indeed: *Bombylius* is led entirely by optical signals. This enables us to "fetch" the animal, that is, to guide it to the spot where we want to do our *Muscari* experiments, simply by putting out a welcome mat consisting of separate, indigo-colored pieces of paper. The glass tubes over the *Muscari* inflorescences do not stop *Bombylius* from trying to approach the latter in a straight line. Had the odor been decisive, the animals would of course have gone to the open base of the tubes.

As soon as we use a yellow glass tube instead of a colorless one, or as soon as yellow paper is introduced into the glass tube to conceal the natural blue color, the attempted visits cease. This probably means that *Bombylius*, in the period before we started our experiments, has trained himself to visit only blue flowers. It has already established a firm tie with them or, to put it in a more vulgar way, is "going steady" with *Muscari*. Blue papers scattered among gray ones in a checkerboard pattern will also receive visits, and the same is true for blue imitation flowers placed among the very fragrant yellow flowers of *Bunias erucago*, a plant belonging to the mustard family.

Does all this mean that the smell has no significance at all? No, we should certainly not draw that conclusion, for there is a profound difference between an arrow-like approach from a distance and a real visit to a flower. In the latter case, *Bombylius* will, with its four anterior legs, grab hold of the white margin of the flower tube, vibrating all the while with its wings in true hummingbird fashion. It inserts its narrow proboscis in the flower and sucks up the drop of nectar that is hidden so deeply in it. We know that many, if not all, flies possess a sense of smell, and it would indeed be most amazing if, in all the operations we described, *Bombylius* would not be impressed by the *Muscari* odor. It is therefore perfectly reasonable to assume that the latter also plays a role in the establishment of steady relations between the animal and *Muscari*. Another possibility which perhaps has not received enough attention is that a whole field of flowering grape-hyacinths will perfume a large volume of air, inducing in *Bombylius* a general food-seeking mood; that is, a willingness to fly *linea recta* to blue objects.

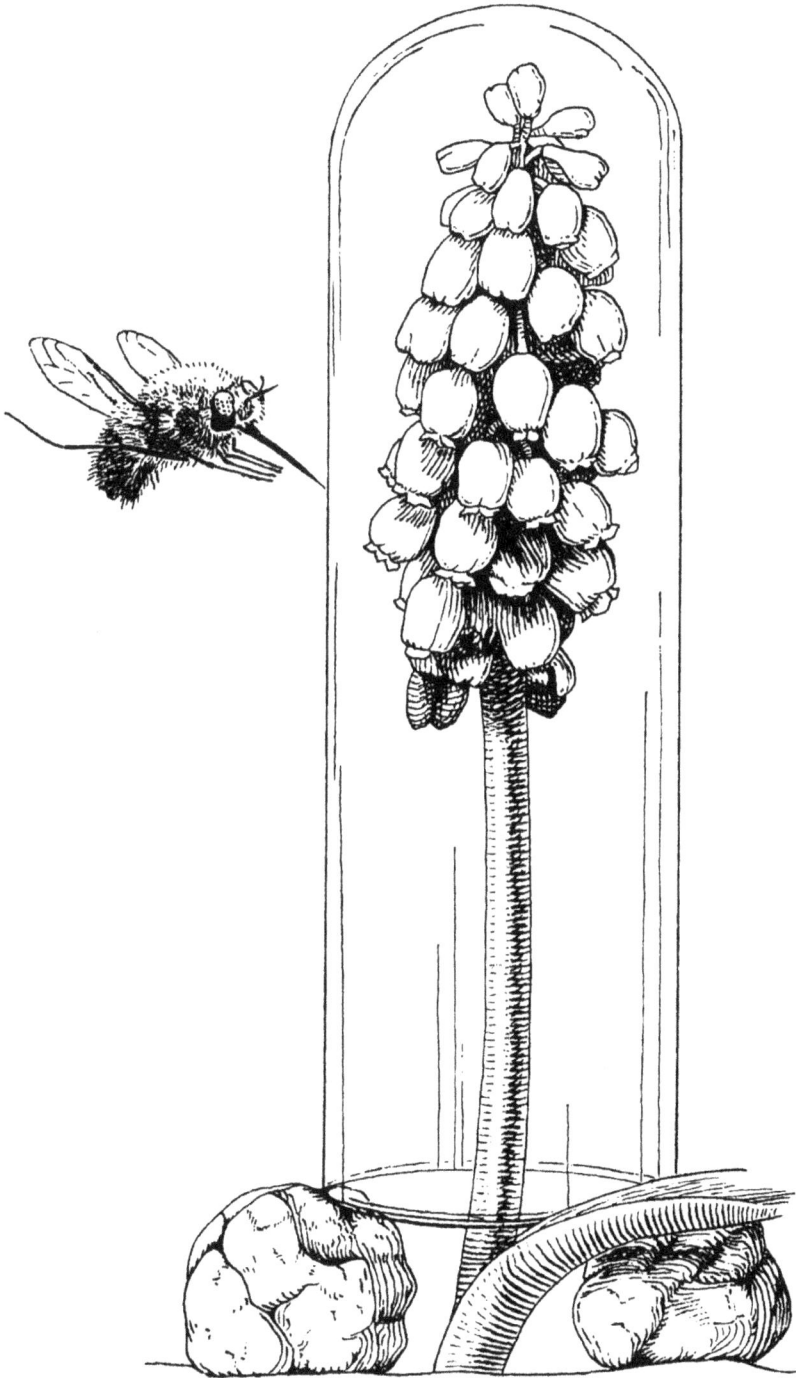

Figure 8. Beefly (*Bombylius*) trying to visit a flowering plant of the blue grape-hyacinth (*Muscari*) that has been surrounded by a wide glass tube. The experiment shows that the insect is guided by visual signals, in this case by the blue color. If it were attracted by the smell, it would try to crawl into the tube at the base where the odor escapes.

After the bees and *Bombylius*, a few words about moths. Some people will be amazed to hear that color vision is important in these animals. Are not moths creatures that are on the wing during the night, and would it not be sufficient if they were guided only by light-and-dark effects or patterns? Well, this point is debatable. There are, after all, many moths that fly during the daytime in bright sunlight.

Knoll has studied one of these rather closely. *Macroglossa stellatarum*, as it is technically known, is sometimes also referred to as "carp's tail" because its rear end is flattened out in a rather conspicuous way. Of course, "whale's tail" would (theoretically) be a better name, since the flat part is horizontal and not vertical, like the tail of a fish. On the other hand, who are we to give such a heavy-sounding name to such a wonderfully fleet creature?

Although red-blind, *Macroglossa* has a well-developed color sense. It recognizes blue and yellow as definite colors, and there is good evidence that the females distinguish green as a color, too, when they deposit their eggs on plants. Just like bees, though, these moths make no distinction among blue, violet, and purple.

*Herse convolvuli*, the European bindweed hawkmoth, is a good example of a moth active late in the evening. It shows an uncanny ability to distinguish different colors in almost complete darkness, leaving us poor humans far behind in this respect. Still, we cannot ignore the fact that the flowers pollinated by hawkmoths are practically always white, light yellow (like some honeysuckles), or pale purplish pink (like soapwort flowers). Simple contrasts in light intensity may therefore play a role, too. It can indeed be shown that white objects of a certain size exert their greatest attractiveness, from the hawkmoth's point of view, when they are placed against a very dark background.

As to the butterflies, I have already mentioned that some 20 years ago I investigated the color sense of the grayling. It turned out to be very much like that of *Bombylius* and honeybees; a strong preference for yellows and blues was noticeable. The same was true for various other meadowbrowns and also for the tortoise-shell and its close relatives, the so-called nymphalid butterflies (see Fig. 6). However, had we used swallowtails and whites for our experiments, the results would have been different. Dora Ilse, who has done the best and most complete work on color vision in butterflies, could show that these (the swallowtails and the whites) very definitely are *not* red-blind, like bees. We have already had the opportunity to point out that there are indeed typical "butterfly flowers" of a pure red color; they must be pollinated by these white butterflies and swallowtails. Ilse could also show that butterflies will be attracted to colors even when they are newly hatched and have never before encountered a flower. We are dealing with an inborn reaction, and a very strong one at that.

In order to get an idea of the magnetic effect which certain colors have on butterflies, it suffices to scatter a few pieces of yellow paper on a lawn covered with dandelion flowers and observe what the visiting cabbagewhites will do.

Smell does not seem to attract butterflies from a distance, but it has to be admitted that there are some exceptions to this rule. In the Mediterranean region and Africa, for instance, we find the beautiful *Charaxes* butterflies that would be difficult to distinguish from swallowtails if it were not for the fact that they possess a *double* set of "tails" on each hind wing. These animals are attracted by fallen fruit, and it is fun to watch how they approach their food source—not in a straight line, like whites and tortoise-shells will do in the case of flowers, but in zigzag fashion, going "upstream" against the wind that brings them the odor-message. Usually, they will sit down when they are still a few inches from the food source, and will cover the last little stretch on foot, with uncoiled proboscis, drumming all the while with their antennae. *Charaxes*, then, is considered to be a typical "nose butterfly" (with the understanding that his nose is in his antennae, of course). However, let us always keep in mind that there is no really sharp distinction between this type and "eye butterflies" such as the tortoise-shell. And even the fact that butterflies can be attracted by certain smells when these are not connected with colors does not mean that colors are unimportant to them.

It would not be difficult to elaborate further on the topic of color sense by recalling Hans Kugler's experiments on bumblebees, those which Fritz Schremmer did on yellowjackets, and so on. However, they all point in the same direction, and I therefore propose to start considering things from the standpoint of the flowers. After having devoted so much time to the color vision of animals, this is no more than fair.

# 4

# A Way To Paint

How does Nature achieve the desired color effects in flowers? Has she been applying, for millions of years, principles which human artists recognized only recently? What are the equivalents, in a flower, of the pigments which a human artist has on his palette? In what form are they available and how does Nature mix her colors?

In order to answer all these questions, we must keep firmly in mind that in the complicated organism which we call "the higher plant" we are dealing with a large group of microscopically small "repeating units," the cells. A typical, living plant cell is somewhat like a cardboard box, heavily lacquered on the inside with a colorless lacquer, and filled completely with water or with a solution of some dye in water. The box itself can be compared with the firm wall of the plant cell, which consists chiefly of cellulose. The lacquer is the thin layer of colorless protoplasm covering the cell wall on the inside. In it, we can have various types of small, colored bodies. Finally, the water or the colored solution which fills the box can be compared with the fluid which occupies the center of a plant cell, forming the so-called "vacuole."

Dissolved in the vacuole-sap of plant cells, we often find pigments which belong to the class of the *anthocyanins*. The name is derived from two Greek words: *kyanos*, which means "blue," and *anthos*, the word for "flower." Actually, the name is a little misleading because anthocyanins range in color all the way from blue through purple to red, vermilion, and scarlet, depending on the particular anthocyanin we are dealing with and also on the acidity of the cell sap. If this sap is very acid, the chances are that the anthocyanin will appear red; if it is more alkaline, then the pigment will be blue. It is interesting to try this out by extracting anthocyanin from red flowers, such as geraniums, which can be done by simply boiling these in water and adding acid or alkali.

The name "anthocyanins" is also misleading because these pigments are found not only in flowers but also in young sprouts and in leaves, sometimes even in roots. Thus, anthocyanins are responsible for the color of red cabbage and red beets and contribute heavily to the flaming foliage hues which make the Indian summers of New England so famous. Yet it is in flowers that we find them with the greatest regularity and in their greatest beauty—we might, for example, mention poppies, red roses, violets, and blue cornflowers.

Next to the anthocyanin pigments, chlorophyll must be mentioned. This amazing green substance makes it possible for the higher plants to harness the energy contained in the rays of the sun and to build up sugars and other materials from the carbon dioxide gas found in the atmosphere. Unlike anthocyanin, chlorophyll is not present in the form of a solution. In combination with many other substances it forms certain green granules, the chloroplasts, which we find distributed in fairly large numbers in the protoplasm of each leaf cell. These green bodies can be seen in a beautiful way by simply putting a living moss leaf, which is very thin, under a microscope. They are found in all leaves capable of forming sugar in the light and also—and this is why we are interested in them here—in some flowers.

The orange and yellow pigments on Nature's palette are the carotenoids. In many cases, these are found in a pure and free state, as in carrots where, under the microscope, we see crystals of carotene in the shape of thick needles and platelets. Even more beautiful carotene crystals can be observed in the orange parts of the famous bird-of-paradise flower, *Strelitzia*. Here they appear as bundles of gently curved, slender and sharp, somewhat flattened needles. The yellow to orange color of California poppies and of many other flowers is likewise due to the presence of carotenoids. In other cases, the impression of yellowness is caused by colored oil droplets, or by the presence of those pigments which we call "flavones"—compounds related to the anthocyanins. Blue or red anthocyanins, green chlorophyll, orange or yellow carotenoids, and flavones, together, cover almost the whole color range we see in a rainbow.

But how about white? one might ask, for there are many flowers in nature which have this color, such as daisies and certain chrysanthemums. Again it is well to remember that "white" is not a color in the real sense of the word. We should say, rather, that the white *appearance* of milk, paper, and porcelain is brought about by the fact that light is scattered, reflected in all directions, when it hits certain small particles distributed in these materials. It is known that in white paper the tiny air spaces between the fibers are largely responsible for the effect. When they are eliminated by filling them up with a fluid, the paper becomes almost completely "colorless" and transparent. This yields a valuable clue. We introduce the white petals of flowers into a tube which is partly filled with water, the air is sucked out with a vacuum pump, and then the original pressure is restored. The petals behave just as the paper did; they lose their whiteness and become quite transparent. This proves that the air in the so-called "intercellular spaces" of the petals, which in our vacuum experiment was replaced by water, plays a very important role, too. Air takes the place which a white paint would have on the palette of a painter. Also—and this is equally important—it can fulfill the role which white paper plays for the artist who produces a water color. We all know what a tremendous difference in character there is between a good water color and an oil painting. The main reason for this is that the white paper reflects the light which reaches it through the thin layer of paint laid upon it. The colors therefore seem to vibrate with light. The same can be said for some flower colors.

From our last comparison, it can be seen that the mere presence of a pigment does not always tell us much, because its effect depends on additional factors. Sometimes, for instance, we can add

Figure 9. Sections through different petals of *Zinnia*. At least three factors contribute to the color: pink or red anthocyanin, dissolved in the cell sap; chlorophyll, present in the discrete green bodies called chloroplasts; and carotene, also present in the form of discrete bodies which are orange in color. The overall color effect depends on the presence of one, two, or all three of the pigments, as indicated in the drawings.

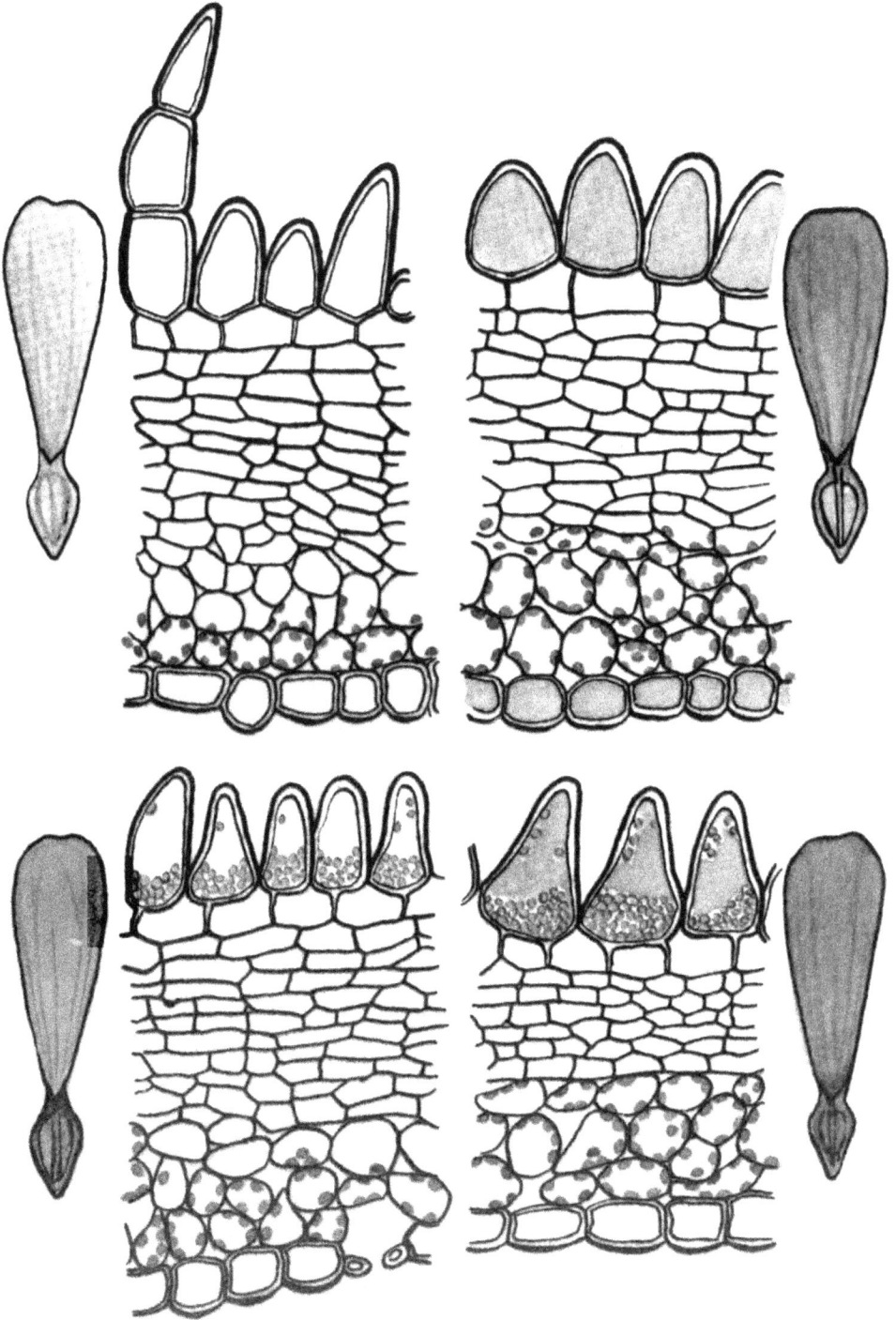

Figure 9

a particular gloss by putting on a layer of varnish. In certain types of cloth we can produce a velvety sheen by a special arrangement of the threads. All these little tricks can also be found in the plant kingdom—and others to boot. But probably the best approach is to present a few striking examples, hoping that this will encourage the reader to investigate many other cases for himself.

First of all, to get a good general idea of the way in which the various plant pigments can cooperate in creating colors, I recommend buying a bunch of good, old-fashioned *Zinnia* flowers. There are varieties where one finds flowers of different colors, red, orange, or greenish, on one and the same plant. Our colored pictures (Fig. 9) show how this can happen. After what has already been said, I do not think it is very difficult to interpret them.

Next, we must consider flowers which apply the principle of "pointillism," made famous by Georges Seurat. Instead of mixing colors on his palette, this artist put tiny dots of the various pure colors side by side on the canvas. He realized that, at a certain distance, the human eye can no longer distinguish these as separate units, so that we get a "mixed" impression. When we examine the outside of the petals of certain red garden poppies or tulips, we notice that some epidermis cells are red and others are blue, due to the presence of differently colored anthocyanins in the vacuoles. The total impression is that of a uniform, violet color (Fig. 10).

When cells of different color are not found side by side, but in layers (one type over the other), we get a different effect. Perhaps the ideal objects for demonstrating this situation are not flowers but the big leaves of a common greenhouse plant, *Rhoeo discolor*, in the group of the spiderworts. Each *Rhoeo* leaf is rather thick, with many layers of cells which are either colorless or green; the only exception is formed by the cells of the skin layer on the underside which lack chlorophyll entirely but contain in their vacuoles a solution of purplish-red anthocyanin. Held up against the light, a *Rhoeo* leaf will give the appearance of almost complete blackness. Not much reflection is needed on the part of the observer to understand why this must be so. The green upper sheets of cells in the leaf act as a color filter, allowing only green rays to go through, and all the other colors represented in white light are eliminated by absorption. However, since red and green are complementary colors, the green rays will, in turn, be absorbed almost completely by the red skin layer of the underside, and there simply is no light at all left to reach the observer's eye.

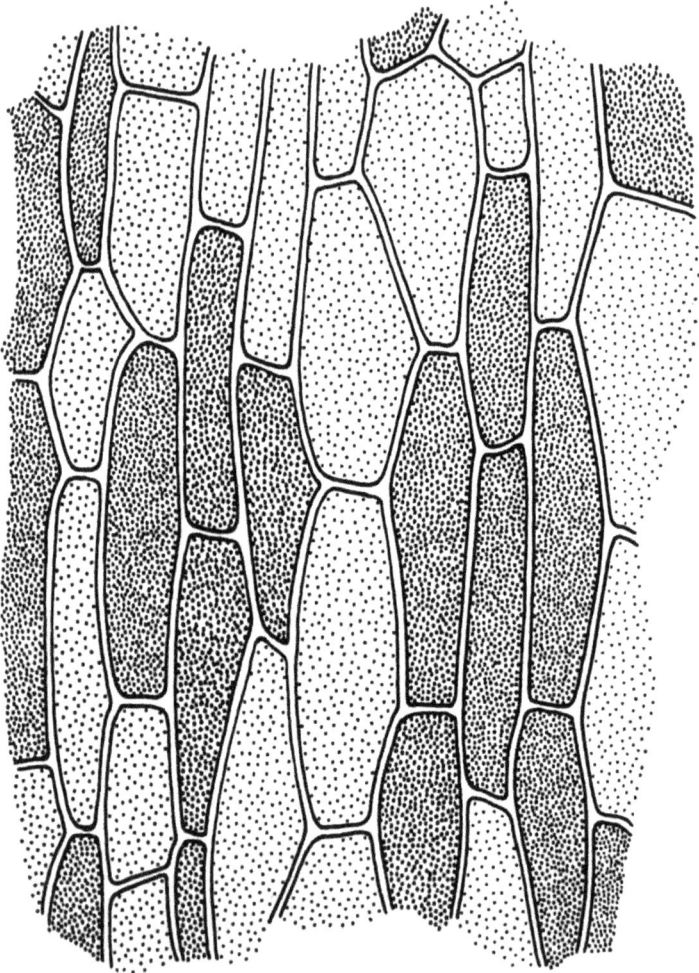

Figure 10. The principle of pointillism, encountered in the epidermis cells of certain tulip petals. A crazy-quilt pattern of completely blue and completely red cells produces an over-all effect of purple.

The black patches at the base of red poppy and tulip petals demonstrate exactly the same principle. No black substance is present in the tissues, but layers of blue-colored cells underlie an epidermis of a deep-purple tone. How important it is for a flower to have such dark patches will become clear in the next chapter, where it will be shown that they often act as honey-guides.

A buttercup flower provides a good example of the water-color effect (Fig. 11). The yellow color is caused by the presence of an oily substance in the epidermis cells. Under the epidermis, we find a layer of prismatic cells so chock-full of starch granules that it presents itself to us as a dazzling white sheet when dissected out. Beyond any doubt, this light-reflecting layer is comparable with the white paper used by the water-color artist. Its great value in producing brilliance of color can be demonstrated on the spot by simple observation of the base of the buttercup petals, where the

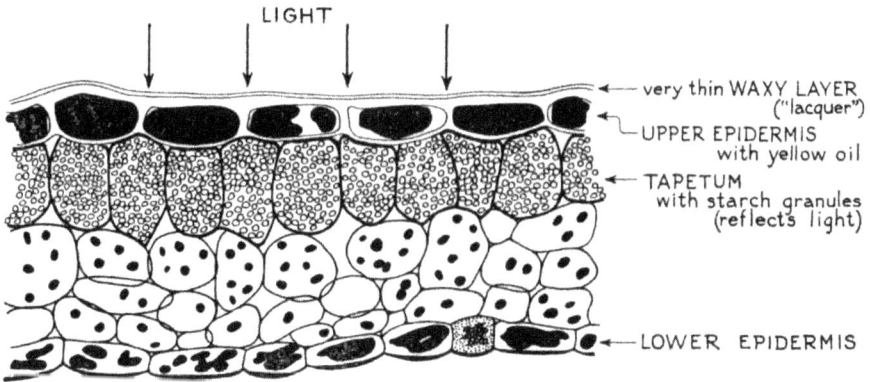

Figure 11. The so-called "water-color effect" in a buttercup petal. Light, going through the epidermis with its yellow oil, is reflected by a special layer of cells chock-full of starch granules. This gives the petals the same sort of luminosity we find in a good water-color painting, where the white paper with its numerous air spaces acts as the reflector.

white sheet is lacking. This part is indeed much duller in appearance.

Some flowers are able to produce the particularly warm colors and the sheen of velvet because their epidermis cells, instead of forming a smooth, flat surface like a tile floor, bulge out so that steep little hills are formed. These can legitimately be compared with the many short bundles of hairs which in velvet stick out at right angles with the surface and which are responsible for the fabric's peculiar light and color effects. One of the most beautiful objects for a demonstration of the velvet effect is the *Gloxinia* flower, although snapdragons, pansies, and some primroses are also excellent for the purpose.

In summary, we can say that it would be very hard to find any principle or trick used by the human painter in the application of his colors which does not have its counterpart in nature. Thus, the mixing of colors on a palette is simulated by the simultaneous occurrence, in certain plant cells, of pigments such as chlorophyll, carotene, and anthocyanin. Pointillism as well as the superposition of variously colored layers is found in flowers; the use of varnish seems to be about as common as the creation of the special surface textures that cause the sheen of velvet. Even those who claim superiority of water colors over oil paintings can find support for their belief in the plant kingdom. All in all, it seems far from unreasonable to insist that the study of color effects in flowers, a delight for both the senses and the mind, be brought to the attention of young artists. Had more emphasis been laid on this subject in the past, it might have been possible to avoid the ridicule and humiliation to which some of our best painters, such as Seurat, have been subjected in their lifetime.

# 5

# *Signposts of All Sorts*

In Chapter 2, we described how Sprengel, in his study of forget-me-not flowers, was struck by the pretty yellow ring around the entrance of the flower tube, and how he thought of it as a "nectar-guide." We mentioned that, afterward, he found similar "signposts" in many other flowers. True, he did not do any real experiments to prove that his idea was right, but at least he made a number of shrewd observations. For instance, he was the first to point out that night-blooming species such as certain evening-primroses, honeysuckles, and silenes lack honey-guides. This is exactly what one would expect, for after all it is not very likely that color contrasts would be of any significance to the moths which pollinate the flowers of these plants in the semidarkness of the evening. Let us add, at this point, that it cannot be just happenstance that these flowers are almost always white, pale yellow, or light salmon-colored.

In different species, Nature has used different means to point the way to the hidden nectar (see Fig. 4). Often, the honey-guide is nothing more than a patch of color forming a sharp contrast with the corolla, as in Sprengel's forget-me-not. Think, for instance, of cowslips and toadflax (orange on yellow), certain irises (yellow on purple), and of blue-eyed grass (yellow on blue). We have already seen how von Frisch demonstrated that there are only four broad "bee colors": yellow (which includes orange), blue-green, blue, and ultraviolet. This being so, it is certainly worth while to examine very carefully the various color combinations which we find in flowers with honey-guides, and this is exactly what von Frisch did. All in all, he examined no less than 94 species, and indeed, the commonest combinations were those that could be most easily distinguished by bees, namely yellow and blue, yellow and purple, orange and blue, and yellow and violet. Again, this cannot possibly be just a coincidence.

There are other cases where the path to the nectar is indicated by a system of stripes. This we see, for instance, in Scotch broom (orange-red stripes on a yellow background), purple loosestrife (dark purple on light purple), and *Iris pseudacorus* (red lines on a yellow background). In still other instances, there are groups or streaks of small dots, good examples being the flowers of foxglove and jewelweed, and yellow monkey-flowers. Some violets and pansies are interesting in that they show a combination of the different principles. Thus, it is not uncommon to find a pansy flower possessing, for a honey-guide, a yellow patch which stands out clearly enough against the blue or violet corolla, but which in addition is striped—for good measure, so to speak.

There are also some plants in which the color of the honey-guide patch changes with time. Among the prettiest examples, undoubtedly, is horse chestnut (Fig. 12). Each of its two white upper petals has a color spot which is yellow to begin with, but which gradually turns through orange to crimson. Later on, we shall see that this change has something to do with the secretion of the nectar, and that it provides the bees and bumblebees with very valuable information.

However, can it be shown by really conclusive experiments that these color patches, dots and lines, described so carefully by Sprengel and others, do indeed play the role of signposts? Many are the people who have pooh-poohed the whole idea. Adolf Engler, for instance, a scientist who was interested in saxifragas, has pointed out that there are some species in this group of plants which have beautiful stripes or dots in their flowers, while in others they are completely lacking. If one assumes that the structure of the flowers is pretty much the same, no matter which saxifraga one chooses, then we must indeed agree with Engler's conclusion that the presence of the lines or dots is meaningless, a whim of nature, and nothing more. But, as Hermann Müller has pointed out in

Figure 12. Inflorescence of horse chestnut, showing the change in color of the honey-guide with age.

Figure 12

his famous work on the flowers of the Alps, this just is not so. If one really scrutinizes the various *Saxifraga* flowers, it turns out that there are all sorts of situations, ranging from the cases with completely exposed nectar to those where it is completely hidden— and the better concealed the nectar is, the better developed is the system of lines or dots pointing the way to it. So, instead of being an argument *against* the signpost idea, the situation in the various saxifragas now is a very powerful one in its favor.

Still, you can argue that this is a case of "indirect evidence." It would be much more convincing if one could prove by simple, straightforward experiments, or by direct observation, that insects are influenced by the sort of patterns which are found in honey-guides. Therefore, let me tell you something about the experiments which Niko Tinbergen and his students, some 25 years ago, did on the homing behavior of a certain wasp, *Philanthus triangulum,* an insect which we used to call the beewolf, because it catches and paralyzes honeybees to serve as food for its offspring. You may well wonder what this has to do with honey-guides, but the connection will soon become clear.

The homes of the *Philanthus* females are to be found on sandy plains, hundreds or even thousands of them close together, at least in favorable years. Each female digs her own nest, almost like a dog would do it: the sand which she kicks backward between her hind legs drops down to form a light-colored patch about the size of your hand, standing out clearly against the undisturbed, dark area around it. All those hundreds of *Philanthus* nests with their sand patches look so much alike that it is very hard to understand how in the world the female *Philanthus* can ever find her own home when she returns with a captured bee from a successful hunt. Could it be that she uses as beacons the small twigs, pine cones, pebbles, and clumps of grass or heather which one invariably finds near the nest?

In order to find out, one can arrange pine cones in a circle around the nest hole in periods of bad weather, when *Philanthus* is forced to stay indoors for days on end (Fig. 13). When the female finally emerges, she will first of all carry out a short orientation flight, during which she seems to memorize the situation around the nest. Then, when the insect is away hunting, we move the circle of pine cones aside, so that the real hole and its sand patch are no longer surrounded. When the wasp comes home with her prey, she will almost invariably ignore her real nest and will alight in the middle of the circle of cones. It is possible to let the insect choose a number of times by scaring it away temporarily with a wave of the hand—and almost always she will go for the cone ring and will start looking for the nest hole, which of course is not there. Eventually, she will find her real nest, but only by accident, and only if the pine cones are not too far away from it. This experiment shows conclusively that the female beewolf is indeed guided by "beacons" in the vicinity of her nest.

The next question is how various beacons compare. Is a pebble more important to the wasp than a pine cone, or is it the other way around? Again, this can be tested by a simple ring experiment. This time we allow the wasp to memorize a situation wherein she has, around the nest, a circle in which, say, eight pine cones alternate with eight pebbles of about the same size. While the wasp is away hunting, we form two new rings out of the old one—one which has only pine cones, and another which contains only the pebbles (Fig. 13). The returning insect chooses the pine cones.

If we try to define the difference in appearance between a pine cone and a pebble, we must say that the former has a well-broken outline; it is the lack of "brokenness" in the pebble that must give it a certain disadvantage as a beacon or landmark. The experiment can be varied in a number of ways. For instance, the wasp can be given a choice between a plain black ring and a black-and-white checkered one; it will select the latter. Smaller checks are preferred to larger ones. Again and again, we find that the insects have a preference for *broken patterns*.

Of course we have to admit that it is a very far cry from the homing behavior of a beewolf to the nectar-gathering behavior of a bee! Did not we ourselves warn against hasty conclusions when we discussed the color vision of the graylings? Yet, it is a comforting thought that at least under certain conditions wasps pay heed to the differences between broken and unbroken patterns and prefer the former. If we can show that they also do this in visits to flowers, then the presence of groups and streaks of dots in the latter must be very meaningful indeed. Unfortunately, no really good experiments on wasps and flowers have been carried out as yet, but there are some interesting ones by von Frisch and his associates

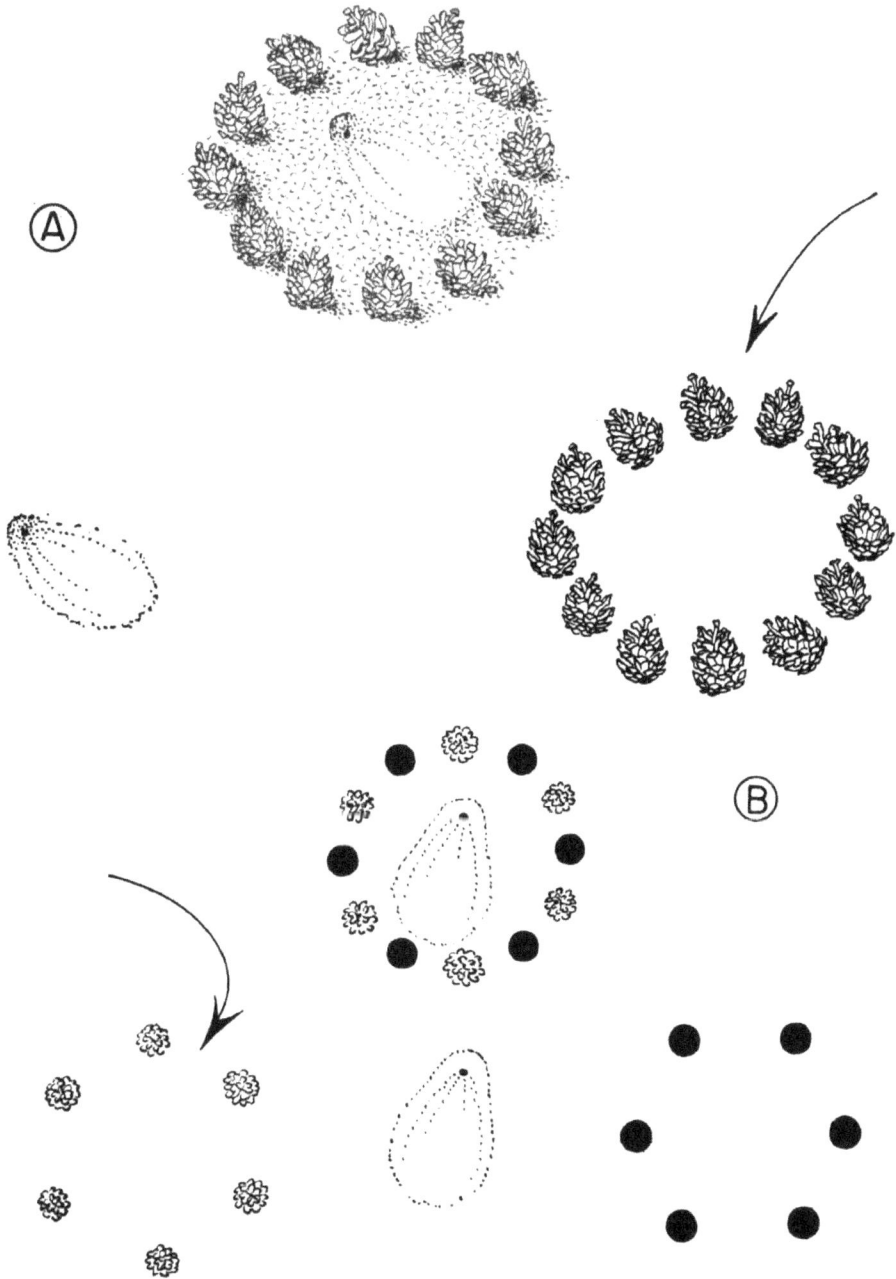

Figure 13. Experiments on the orientation of the beewolf, *Philanthus triangu-lum.* A, trained to a circle of pine cones surrounding the nest (upper drawing), the homing beewolf will move toward the circle of pine cones when the circle is offered separate from the nest (lower drawings). B, trained to a circle in which round, smooth objects alternate with highly "articulated" ones (upper drawing), the homing beewolf will prefer the circle of articulated objects over the round, smooth ones when two separate circles (lower drawing) are offered. (After N. Tinbergen.)

on bees and bumblebees in search of food sources provided by man. The results are in complete agreement with those of Tinbergen— "richness of outline" is of paramount importance. Thus, although it does not seem possible for a honeybee to learn the difference between the shapes in the upper row of Figure 14, or between those of the lower row, for that matter, it will readily distinguish between any of the figures in the lower row and any of those in the upper.

Still, no matter how good these experiments by von Frisch on the recognition of shape may be, nobody would ever claim that

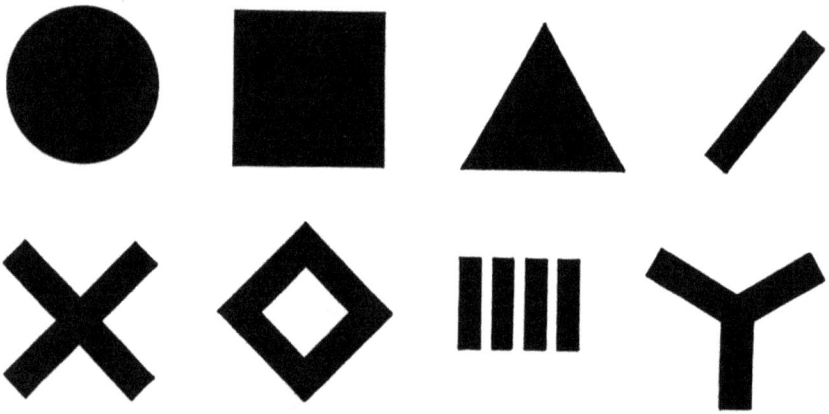

Figure 14. Different shapes used by von Frisch to study the perception of form by honeybees. Any item in the upper row can be distinguished from any one in the lower row, but the animals cannot learn to differentiate between the objects in each row. What counts is the degree of brokenness. (Redrawn, by permission, from Karl von Frisch, *Bees: Their Vision, Chemical Senses, and Language.* Cornell University Press, 1956.)

they have solved the honey-guide problem completely. They were concerned, rather, with the approach to certain big objects *from a considerable distance.* A study of the structure of a bee's compound eye reveals that, essentially, it is a group consisting of a rather small number of individual units. For that reason, it "translates" reality into an image in a rather crude fashion, with much loss of detail. It would take us too far afield to explain this thoroughly. Let us, therefore, only say that there is some resemblance to television, which also depends on a number of units, each responsible for the "translation" of a small part of the real thing. The higher the number of units, the better the picture, and vice versa. On the basis of this, it must be considered very doubtful that honey-guides, which are very limited in size, can be distinguished by insects until they are very close. This last consideration, by the way, is also the main objection which we have against the conclusion which Hans Kugler, a bumblebee expert, has drawn from the many experiments which he has carried out on the function of honey-guides. His idea seems to be that these structures make the flower as a whole more attractive to insects, so that it will receive *more* visits and not necessarily better-directed ones. This is an amazing conclusion, especially when we keep in mind that Fritz Knoll, in a series of experiments for which he is rightly famous, had already demonstrated the directing role of honey-guides in the case of our old friend, the hummingbird-hawkmoth *Macroglossa stellatarum*.

Earlier (page 26) we mentioned that *Macroglossa* flies in bright daylight. It is obviously quite visual in its responses to food sources. One of the flowers on which it can often be observed is our common toadflax or butter-and-eggs, a yellow wildflower which shows its close relationship to the snapdragons of our gardens by the possession of a long, nectar-containing spur and a slit-shaped flower entrance, which in this case is indicated by a bright-orange patch on the lower lip. Knoll managed to keep captive macroglossas in good shape by allowing them to fly around in a large cage and to feed from flower models which contained sugar water. This left their tongues wet and sticky. He now set up a flowering toadflax plant in a vertical position, pressed tightly between two glass plates. Hovering in front of this dummy, the hawkmoths would try to stick their tongues into the flowers, and of course they would each time touch the glass and leave a small, sticky smear there. A permanent record of their hits could be obtained by spraying the glass plate afterward with red lead and heating it. The proboscis marks that showed up always coincided with the spots where the honey-guides had been (Fig. 15). To be doubly certain, Knoll cut the orange honey-guide out and attached it to various parts of the flower. Even then, the moths would always aim at it. Finally, he made models of honey-guide flowers, in the shape of an ellipse on which the honey-guide was indicated by a small circle or a group of smaller dots, and studied the response of *Macroglossa* to these structures. The result, shown in Figure 16 based on Knoll's original illustrations, beautifully demonstrates the guiding effect.

The behavior of the moths toward natural, accessible toadflax flowers further confirms the conclusions reached in these experiments, for Knoll observed that in a group of toadflax plants, or in one inflorescence, a few flowers are sometimes found in which the honey-guides are lacking. The hovering hawkmoths will not stick their tongues into these, or they will give them a try only after they have visited all the other, normal flowers.

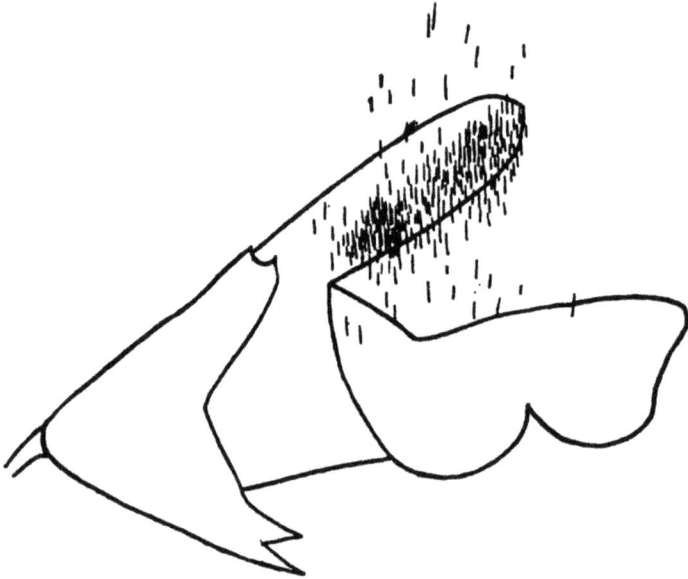

Figure 15. Proboscis marks left on glass by hawkmoths aiming for the honey-guides of certain flowers. (After Fritz Knoll, by permission, Uitgeverij Servire, publishers, The Hague.)

At this point, a brief summary may be in order to show how much was certain as a result of all these investigations and how much remained to be proved.

After Sprengel had postulated the signpost function of honey-guides without giving much experimental proof for his idea, von Frisch demonstrated that the commonest color combinations which honey-guides form with the rest of the flower are indeed those which can be distinguished most easily by bees. Within a group of closely related plants such as the saxifragas, the best honey-guides, in the form of patterns of dots or lines, are indeed found in those species that have the best-concealed nectar. Certain insects such as honeybee and beewolf do show a preference for broken patterns in at least some of their activities. Finally, Knoll showed that hawkmoths which fly in the daytime, such as *Macroglossa,* aim their long tongues at honey-guides of the color-patch type.

However, there was still room for some doubt. Do honey-guides really guide, or do they merely help to attract insects from a certain

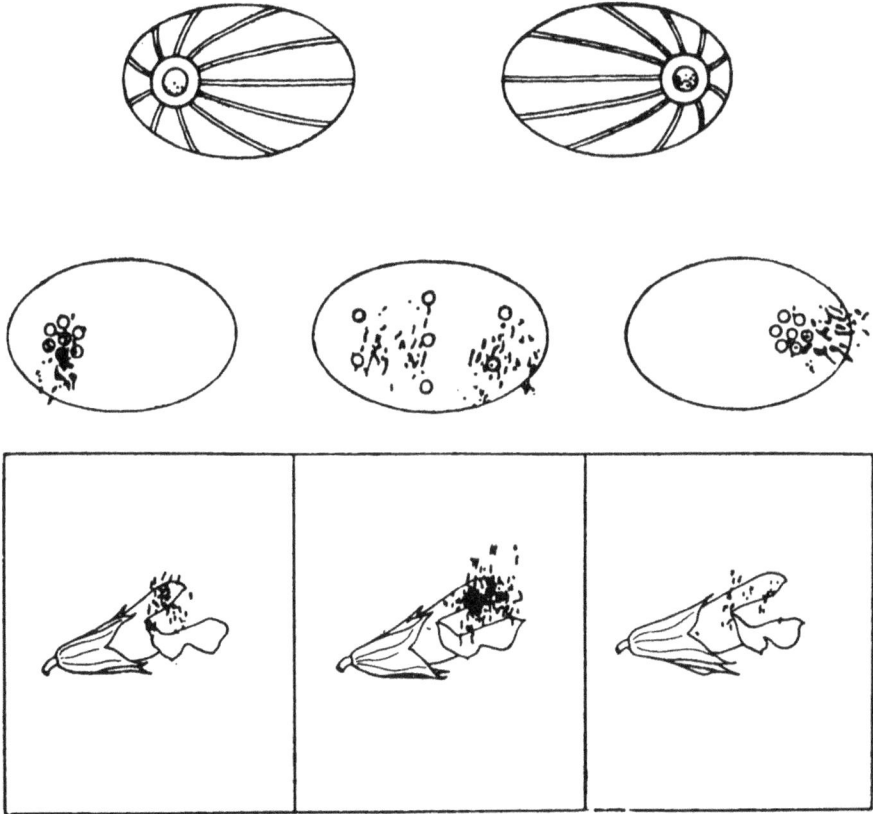

Figure 16. Flower models used by Fritz Knoll in his honey-guide experiments. (By permission, Uitgeverij Servire, publishers, The Hague.)

distance, as Hans Kugler believed on the basis of his experiments with bumblebees?

It stands to reason that the results obtained with the hawkmoth, *Macroglossa,* should never be used to "prove" that Kugler's interpretation of certain experiments on bumblebees was wrong. Clearly, further experimentation was needed. It was furnished by one of Tinbergen's students, Aubrey Manning.

Manning accommodated colonies of bumblebees, which he had dug up in the field, in nestboxes and did his experiments in a big gauze cage of $8 \times 4 \times 4$ feet, where many of his animals felt quite at ease and performed well. In order to determine the maximum distance at which a honey-guide can be seen, bumblebees were accustomed to receiving sugar water on two blue, circular disks of 4-inch diameter, one plain and one with a honey-guide pattern of yellow lines like the spokes of a wheel. Both models were then offered without sugar water and the course followed by the bumble-bees was watched. The number of responses was about the same for each model. At a distance of about 2 feet from them, it was clear at which of the two models the animals were aiming. That the bumblebees could see any difference between the two disks at this distance must be considered very unlikely. However, once they were close to a model, it was quite evident that they saw the differ-ence. They dipped down much more often toward the disk with the yellow lines than toward the plain one. In a way, then, it is true that the presence of a honey-guide does make a flower (or a model) more desirable; however, the effect appears only after the bee has been attracted from afar. The honey-guides *do* induce the bees to alight (as indicated by their dipping), but only after the animals have already arrived on the spot!

How important it is for a flower to have the means to guide a bumblebee to its center is demonstrated by some other experiments of Manning, in which a tremendous preference of the animals for the *edge* of models (flowers) came to light. Three plain models of bright-blue paper, roughly equal in size ($6 \times 4$ inches) but dif-ferent in shape, were offered to the animals. One of the models was circular, another star-shaped, and the third one primrose-shaped, which means that it had the form of a star with very broad, blunt arms. In the crucial experiments, these models were offered without food. The bumblebees did not alight on them, but would hover over the paper with an occasional quick dip down from a distance of about an inch. Even those bumblebees which previ-ously, during a certain training period, had learned to expect a dish of sugar water in the middle of the star-shaped model would aim their dips *at the edge*—the line of contrast between the blue paper and the background—four to five times more often than at the center. This agrees well with the observation that in nature, too, bumblebees usually alight on the edge of large flowers such as thistles. However, when the experiment was repeated with a blue star with a bright-yellow line on each "petal" pointing toward the center, the boot was on the other foot. Bumblebees who had learned to expect food in that center would now pay many more visits to the middle of the model than to the edge. A circular yellow dot in the middle, which gave the model some resemblance to a huge forget-me-not flower, likewise acted as a very powerful guide. Finally, Manning could show that a blue model with a darker-blue center also elicited more responses than a plain one.

Manning's final conclusion is that foraging bumblebees always begin to fly toward a flower's edge. If there is a honey-guide, they soon find their way to the center; in the absence of a honey-guide, this is much more difficult. On large *Magnolia* flowers, which are honey-guide free, one can see that the bees are indeed at a loss. Again and again they fly to the edges and tips of the petals, and although some of them eventually find the flower's center (after considerable searching!), there are several others which never make it and simply give up. Later on we shall see that *Magnolia* is a beetle flower; one could say that it was not meant to be pollinated by bees.

Summarizing, it is safe to say that, after the old and somewhat "isolated" experiments by Knoll on the hummingbird-hawkmoth, Manning was the first scientist clearly to demonstrate the role of honey-guides. However, he himself has pointed out that in the study of these structures many riddles remain to be solved. Why is it, for instance, that many of the most intricate honey-guides are found on small flowers such as forget-me-nots, eyebright, and speed-well, in spite of the fact that large flowers obviously are more in need of honey-guides than small ones? Do the small flies, which in this case are the main visitors, respond to honey-guides in the same way as bumblebees? And what will the study of "chemical honey-guides" reveal? Thus far, what little work has been done has been concerned only with visual structures. The only exception to this is the investigation of Miss Th. Lex, one of von Frisch's students. She found that honey-guides differ from the rest of the flower not only in color but also, as a rule, in smell. For instance, the very conspicuous honey-guide of the true narcissus, *Narcissus poeticus,* has a fragrance which, to us, is both different from, and stronger than, that of the other flower parts. By training experiments, one can show that bees also notice the difference. The same sort of situation exists in primroses, violets, certain irises, and nasturtiums. In horse chestnut, the smell of the honey-guide is stronger than that of the corolla, but is not otherwise different to us. The quality of the smell, however, changes with the change in color of the honey-guide from yellow to red. Furthermore, Miss Lex could show that in some cases there are honey-guides which are conspicuous *only* by smell and not at all by color! The example of one of the common wild bindweeds, *Convolvulus arvensis,* is very illuminating. In this plant, the flowers are sometimes white and sometimes pink. The honey-guide—five white lines, like the spokes of a wheel (see Fig. 4)—is conspicuous by both color and smell in the pink flowers, but by smell alone in the white ones. For that reason, it is appropriate to close this chapter with the remark that for the time being there is much more to this business of honey-guides than meets the eye.

# 6

# *Unbidden Guests Beware*

In our last two chapters we have seen that flowers would seem to engage in "advertising" their wares, nectar and pollen, with bright colors. They often have an attractive smell. They seem to try to lead would-be visitors to their hidden treasures with the aid of honey-guides. In short, it is almost as if they have rolled out the welcome mat for all comers. But let us be careful in our judgment. The defense against unwanted, harmful visitors is a principle just as important as the attraction of beneficial ones. To see this, let us just compare the performance of efficient pollinators such as bees, hummingbirds, and hawkmoths, all fleet of wing, with that of the slow, crawling ants. In a country the size of prewar Germany alone, honeybees pollinate about 10 trillion flowers in the course of a single summer's day. Since there must be far less than 10 trillion honeybees in Germany—the United States, which is much bigger, has only 90 billion bees in 6 million colonies—it follows that each individual bee takes care of a great many flowers, and the pollen which it carries has a good chance of getting to the right stigma before it has been in the air very long. Thus, it has lost little of its vitality, and as a rule the stigmas are still in a healthy, receptive state, too.

On the other hand, even if ants could be induced to carry pollen from one flower to another, what a long and toilsome journey it would be! The pollen would have a very high chance of getting lost. It might be rubbed off on plant hairs or washed away by rain water, and even if the precious substance did finally arrive at its destination, the stigmas might no longer be receptive. We will

pass in silence over the ugly fact that ants might also keep beneficial, pollinating insects away. In view of all this, it is not amazing
that only one case of ant-pollination has ever been described,
namely that of *Orthocarpus pusillus*, a truly diminutive American
plant. It is clear enough that, in general, flowers must try to keep
ants out. In fact, they do more than that, for not all of their fleet
visitors are equally welcome. A tiny fly, for instance, can steal
nectar without so much as touching the anthers. This sort of pil-
fering cannot simply be shrugged off. Real damage is done, for the
flower is robbed of some of that nectar on which it relied to bribe
some larger and more effective animal. Quite obviously, then, all
animals which do not help in the transfer of pollen because they
lack the right body size or shape or speed must be regarded as
unbidden guests, and the plant must do its utmost to prevent them
from getting at the honey. It is interesting to see how many different principles have been employed in this sly fight between the
plant and the would-be robbers, and how many "lines of defense"
there really are.

In the first place—and I am almost afraid to call this a principle
—there is bribery pure and simple. The plant buys off the unwanted visitors by offering them nectar secreted in the region of
the leaves. Of course, this method works only against the wingless
little marauders that creep up from the ground, but against these,
at least, it is supremely effective. Many of our greenhouse balsams
(*Impatiens* species, relatives of the wild jewelweeds) are excellent
examples. Close examination of the plants reveals two nectar-
producing glands at the foot of each leaf. In a Himalayan balsam,
*I. tricornis*, one of these is so beautifully developed that it has become a fleshy cushion, fused with the base of the leaf as well as
with the surface of the stem, and oozing out nectar from its down-
ward-facing surface. The whole gland is placed in such a way that
insects such as ants which walk up the stem cannot possibly avoid
it. That the trick is effective is shown by the fact that the flowers
are free from ants, whereas the glands have many of them as visitors. That the leaf glands are really there to bribe the insects is
clearly demonstrated by the circumstance that they begin to produce nectar *just at the time when flowering starts*. Otherwise, it
might perhaps be thought that our balsams were in that group of

plants which provides ants with a more regular salary in the form of nectar, in order to keep leaf-eating intruders such as caterpillars, snails, and beetles away.

Next to bribery, we must mention the principle of isolation—the nectar in the flowers is made hard to get at. There are cases where this can hardly be called a problem, because all that is really required of the flowers is a negative effort: they simply fail to roll out the welcome mat. To see this, let us first take a flower which relies on bees or bumblebees for its pollination. Practically always, it offers these animals a landing platform of some sort, or at least something they can hang on to. Likewise, certain Asiatic flowers which are pollinated by birds offer their visitors perches in the form of branches or twigs that are placed in a convenient way. However, a flower pollinated by our American hummingbirds, animals that hover in front of blooms and touch them only very lightly, usually shows a complete absence of landing structures or perches. As a result, other flower-visiting animals which require a perch, such as bees, usually cannot visit a typical hummingbird flower readily or effectively. Even birds are not equal to the task if they are not hummingbirds.

Leendert van der Pijl, an excellent student of the flowers and their friends, whom we shall have the opportunity to mention more often in this narrative, has illustrated this in a very striking way by pointing out what has happened to American bird-flowers which have naturalized in Indonesia. Hummingbirds, being characteristic only of the Americas, are of course lacking in that archipelago. The Asiatic flower birds are unable to get at the nectar in the way Nature intended it to happen. They are completely at a loss, and in most cases they end up by forcing their way into the flowers in a process which can only be described as burglary.

Another case where isolation of the nectar is no problem is that of plants which in their normal life are surrounded by water—water lily, arrowhead, bladderwort, frogbit, flowering rush, and many others. However, there are some land plants, too, that have learned to use water as a barrier. In teasel or *Dipsacus*, for instance, we see that each leaf has its own little basin of fluid at its base. This, we must admit, is a rare case, and many naturalists are not even convinced by it. Much more often, it can be seen that milk juice or latex is used to set up the barrier, or some other sticky or tarry, birdlime-like material. This reminds one of the way in which nurserymen protect their trees. They tie a sticky cloth or rag around the stem or paint the bark with birdlime. Next to milkweed, wild lettuce is the outstanding example. In the scales which envelop the flowerheads of this plant, there are many branching latex tubes, some of them so close to the surface that they actually stick out as tiny hairs. Because of this, they are very easily damaged by the sharp claws of ants. When this happens, the hapless insects are simply glued to the surface by a mass of latex. When they bite at the scales in self-defense, the only result is that their heads also become engulfed in the sticky mess. The harder an ant works to free itself, the worse things get, and it is not at all amazing that the flowerheads are eventually garlanded with the dead bodies of a score of ants.

Sticky secretions on the flower stems or the stalks of inflorescences are sometimes so striking that plants have been named for them. In England, some silenes are referred to as catchflies, and the Latin name *Silene muscipula* for a particular species means the same thing. Some silenes exhibit another interesting feature. Their flowers, open on successive nights, are often fragrant and very successful in attracting moths. In the daytime, however, these same flowers are all shriveled up and odorless, as if they wanted to make themselves as unattractive as possible.

Even this does not exhaust the ingenious devices of that remarkable genus, *Silene*. We get the impression that during the process of evolution some species have hit upon the superbly simple device of making nectar inaccessible by keeping it in the center of a calyx that is inflated to the point of being balloon-shaped. Even those insects which are able to bite a hole in the wall of the balloon are powerless if they do not possess a very long tongue as well. Most bumblebees, to be sure, have such a tongue, and also the strong jaws necessary for biting. As a rule, however, it is just as easy for them to get at the nectar in the regular way.

Many of the pinks are provided with still other safeguards against the onslaughts of bumblebees. They have surrounded the flower tube, with its nectar, with a coat of armor in the form of thick scales which cover one another like the tiles on a roof. Even the strongest bumblebee cannot pierce these, and the only thing left to do for the animal is "to go straight."

It would not be difficult to mention still other devices to bar unwanted visitors. We could speak of the sticky hairs on many flower stems, the hairtufts within certain flowers that may serve as a last defense against winged marauders (if they are not a protection for the ovules, an explanation favored by modern scientists in many cases), and so on. But we must leave something for the reader to discover, too.

If we look at the situation from the viewpoint of the animals, we must come to the amusing conclusion that, in a way, they sometimes bar themselves. For instance, they may lack the proper weight. That this factor can be of paramount importance is demonstrated beautifully by Scotch broom, *Cytisus scoparius* or *Sarothamnus scoparius*. Rumor has it that this plant was introduced into North America by Thomas Jefferson. Be that as it may, it has flourished and spread about as well as the United States has. In western Washington and Oregon, for instance, many hillsides are one gorgeous mass of gold when the Scotch broom is in bloom, usually in April and May. For good measure, beautiful color varieties of Scotch broom and its close relatives are grown in our parks, so that there really is no excuse for not knowing this plant. For its phenomenal success, insects must be held responsible and especially the large bees such as carpenter bees (*Xylocopa*) and bumblebees (*Bombus*), for only these are big and strong enough to trigger properly the explosion mechanism of the flowers. Honeybees have to work very, very hard to get anywhere.

Although the broom flowers usually have dark honey-guide lines, pollen is the only thing they have to offer. This, however, they give in abundance. It is really a pleasure to figure out how the explosion mechanism works. In a young flower which is not yet open, the keel and wings actually form one unit, the keel "hooked in" by a tooth on each side. The column of stamens, plus style, is for the time being kept in a straight position, but it is under teriffic tension. The style, especially, seems to be just waiting for the moment that it can pop out and curl up like a hoop. This moment arrives when a heavy insect alights and begins to push the keel down. Starting at the base, the upper margins of the keel come farther and farther apart, the hidden column can no longer be kept in check, and all of a sudden the five short stamens pop out to powder the visitor's belly with pollen. Immediately afterward, the style and the long stamens follow suit but these curl around until they touch the bee's back (Figs. 17, 18, 19). For this, they turn out to have exactly the right length and curvature. Here, then, is a flower that does not seem to neglect any angle.

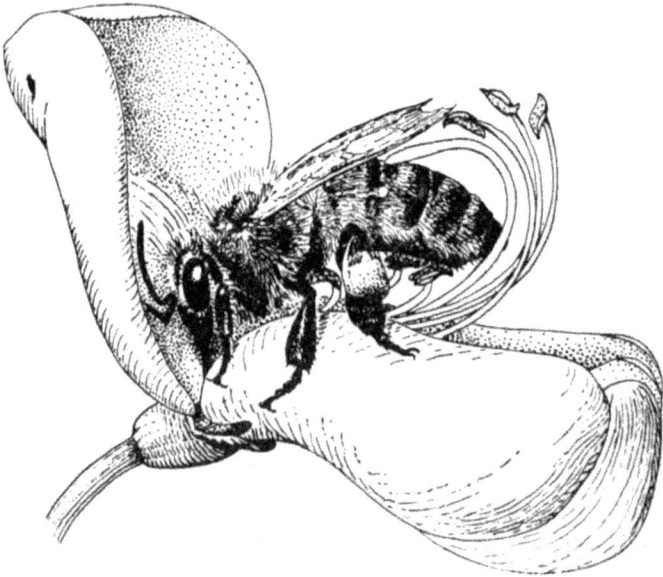

Figure 17. Honeybee pollinating a flower of Scotch broom, *Cytisus scoparius.*
Only heavy insects such as bumblebees and some other bees can set in motion the explosive mechanism for which Scotch broom is famous. In the drawing, some stamens have been left out for the sake of clarity.

By watching the toiling honeybees, one has a fair chance to follow the whole process. However, it is a great deal more exciting and dramatic to wait for the arrival of a bumblebee. In spring most of the bumblebees one sees are the colorful queens that have hibernated (see Chapter 11). They are very active and, best of all, heavy, so that they are infinitely more efficient than the honeybees. Going from one flower to another at a breakneck speed, they cause explosions all over the place, the air is full of tiny pollen clouds, and the whole scene is strongly reminiscent of a Civil War picture or an old western, full of gunsmoke. It is one of the really delightful sights of spring.

Figure 18. Honeybee tripping flower of Scotch broom and being pinned down temporarily as a result.

*Spartium,* a plant that grows very abundantly in the Mediterranean region and sometimes also in American parks, has flowers that behave in exactly the same way as those of Scotch broom.

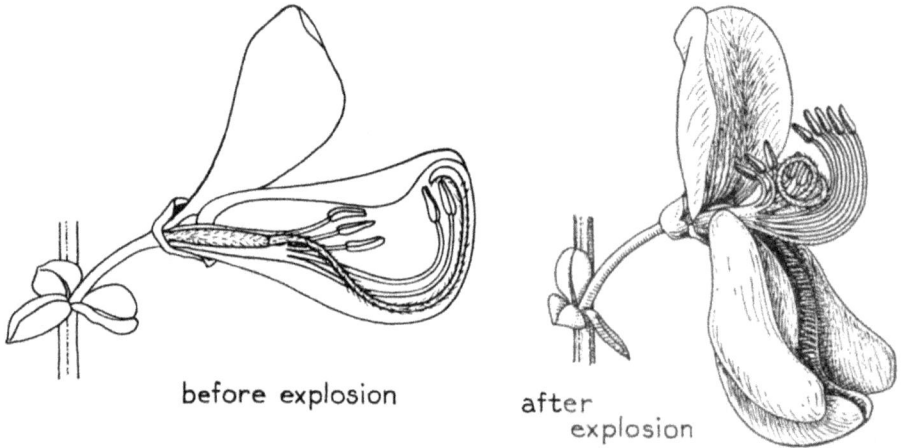

before explosion                    after explosion

Figure 19. Flower of Scotch broom before and after explosion.

# 7

# Burglars and Law-abiding Citizens

What can the unfortunate insect do if it is found wanting in weight? In the case of honeybees and Scotch broom, discussed in the previous chapter, one does not have to watch very long to find the answer. The bees soon learn that the column of stamens will pop out when they stick their head into the slit of the keel near its base and then move toward the tip, forcing the edges of the keel apart. The procedure always pays off handsomely. It can easily be understood why Sprengel was so deeply impressed with the activities of bees and bumblebees and with the wonderful ways in which these animals and certain flowers fit together. Perhaps we should also try to understand his disappointment, nay anger, when he found that there were certain cases in which the animals did not live up to his high standards, cases of what he called burglary and nectar-theft, in which Nature was thwarted in her purpose and no pollination occurred.

Nowadays, we have a much more philosophical attitude about this. We accept it as a most interesting bypassing of the defensive mechanisms which the flowers have set up and which we have just discussed. It is simply the other side of the coin. This burglarizing shows that, occasionally, Nature *does* overshoot her mark; things *do* get stretched to the breaking point and beyond. It is altogether fitting and proper that we discuss burglary here, especially because it may have important consequences for the survival of certain plants, which in turn means that it has a bearing on evolution.

The phenomenon of burglary must have something to do with the tongue lengths of bumblebee species. *Bombus hortorum*, for instance, the "garden bumblebee" of western Europe, possesses a very long tongue—19 to 21 mm in the queens (almost an inch) and 14 to 16 mm in the workers. Therefore, this insect has no trouble in getting the nectar even from *Corydalis* and other long-spurred flowers. Its colleague, *B. terrestris*, however, is by no means so fortunate, having a tongue which is only half that long. Luckily, Nature has provided *B. terrestris* with strong jaws. Confronted with a *Corydalis* or columbine flower full of nectar, it will sit down on a spur, almost like a rider on a horse, and bite two small holes. Then it will stick its proboscis in and "steal" the nectar, if we may borrow Sprengel's terminology here for a moment. In some cases, biting does not even seem to be necessary; the bumblebee will just use its tongue or proboscis as a rapier and thrust it through the wall of the spur to get at the honey. The whole operation takes place on a trial-and-error basis, so that the animal may probe around for quite a while before it has found the right spot. It does not know that spot instinctively and apparently is not guided by a special smell, either.

There are quite a few species of flowers which are thus "victimized" by bumblebees—and also, I am sorry to say, by our highly esteemed honeybee. I hasten to add that the latter usually just takes advantage of the holes already bitten by bumblebees. Let us mention just a few: gentians, soapwort, monkshood, columbine, comfrey (Fig. 20), red clover, bleeding hearts, vetches, beans, blueberries, and mertensias. Mertensias are especially wonderful flowers to watch to see the robber insects, criminals from Sprengel's point of view, in action. They are quite common in many regions of the United States and are very attractive to all kinds of bees and wasps (and also to hummingbirds, by the way). In the eastern United States, we find *Mertensia virginica*, which is usually referred to as Virginia bluebell or Virginia cowslip. In the West; the species is *M. sibirica*. I have very happy memories of trips through the Olympic Peninsula, during which I could see the bumblebees perform for me on the flowers of this plant.

It seems that there can be great differences, even between the members of one and the same bumblebee colony, as far as the burglarizing habit is concerned. Some individuals, having successfully visited a number of flowers in the "legitimate" manner, will continue to do so, while others, after a few unsuccessful attempts, will burglarize one flower after another. The striking thing, in the case of our mertensias, is that quite often the small wild bees do not even wait for the buds to open before they force their way in. The same is true for the fly honeysuckle, *Lonicera ciliata*, a species of the northern woodlands. In their haste to reach the delectable nectar, bumblebees often cut the buds into shreds.

How serious the burglarizing habit of bees can be from the point of view of the plants is clearly demonstrated in some parts of

Figure 20. Bumblebee burglarizing a flower of comfrey (*Symphytum offi-cinale*). Another flower has been victimized previously, as indicated by the two small holes near its base. Excessive burglarizing of certain species of flowers, precluding legitimate pollinations, may conceivably lead to extinction of these species.

Holland where *Corydalis*, a frequent victim, very rarely sets seed. Obviously, there are not enough "law-abiding" bumblebees and long-tongued *Anthophora* bees around to visit the *Corydalis* flowers in the normal fashion and to compete with the "burglars." Anton Kerner, after having studied several species of *Corydalis*, monks-hood, and *Silene* in the Alps, has claimed that some of them are becoming extinct as a result of burglary. The logical question is why this did not happen long ago? The answer, and the reason why we can still observe the last phases of the process, must be that these particular plants in the Alps date back to a period during which bumblebees did not visit the regions where the plants grow and the flowers needed protection only from creeping insects. It is fascinating to realize that the simple observation of insects and flowers may thus help to give information on the geological history and the climate in the past of certain regions.

# 8

# A Matter of Timing

One of the properties of honeybees which we have not yet discussed is their phenomenal time sense, described for the first time in 1929 by one of von Frisch's students, Ingeborg Beling. She found that, if bees were fed at a certain time, let us say between 10 and 12 in the morning for a few days, they would for the next few days return to the feeding spot at approximately the same time, even though the food dishes were now kept empty. Our graph (Fig. 21) illustrates a slightly modified experiment of this sort, in which the honeybees were trained to receive breakfast, lunch, and dinner. This turns out to be a little more difficult for the animals, but—as you can see for yourself—their performance can still be called quite good.

Ingeborg Beling found that her bees would continue to do an excellent job if they were kept in a dark room, at a temperature different from that in a normal hive. This shows clearly that the bee's time sense has nothing to do with the position of the sun in the sky. We must think of it as a wonderful built-in time clock.

The following experiment is very revealing. Between feeds, bees were flown from Europe to New York, where they were put in a room exactly like the one they were accustomed to. Of course, the time in New York was five or six hours behind that at home, but the bees did not pay the slightest attention to that fact; they came out for food at the moment which corresponded with their right home-time.

Needless to say, the bees' built-in clock is of the utmost importance in their lives. Many plants do not produce nectar all through the day but only at particular times. A good example is wild

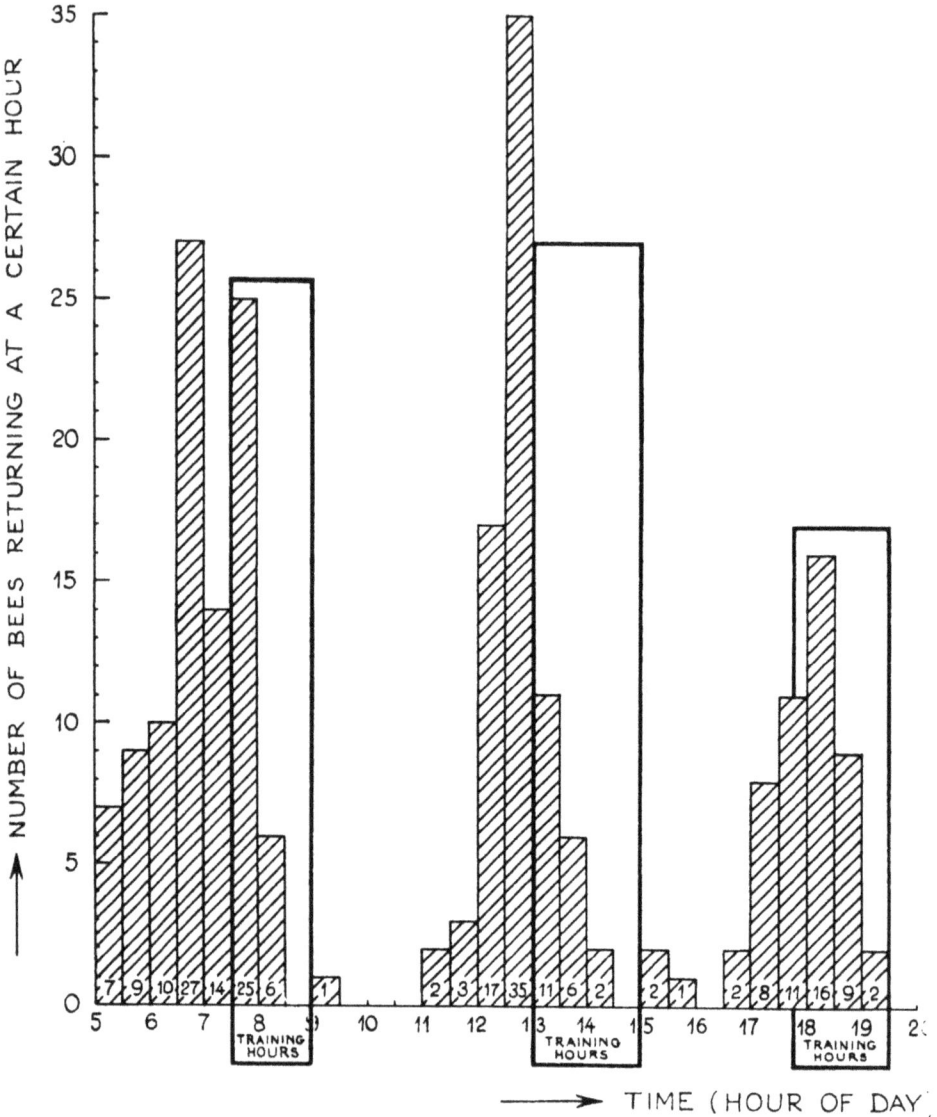

Figure 21. Graph illustrating the time sense of bees. Trained to receive food during three well-defined periods each day, the animals returned at approximately the right time on a number of subsequent days when the food was withheld. (Modified after Ingeborg Beling, by permission, Uitgeverij Servire, publishers, The Hague.)

chicory. Its beautiful light-blue flowers, which in the summertime make our roadsides so gay, yield nectar only between 7 and 12 in the morning. In other plants, there is a peak in nectar production, or in nectar quality, at certain hours. Pollen, too, is in some cases released at very sharply defined times, for instance in mullein, red poppies, and some of our wild roses and bindweeds. The bees obviously learn to recognize these times, for they do not visit the flowers except at the appropriate hour. In this way, they do not waste their foraging efforts. Should you feel a little bit dubious about our story, then it will be illuminating to count the number of honeybees which you notice on one kind of flower at various hours of the day. For wild mustard and certain dandelions, you will find that traffic is heaviest around 9 A.M. The rush hour for blue cornflowers is at 11 A.M.; for red clover, fireweed, and marjoram around 1 P.M.; for viper's bugloss and bachelor's-buttons, finally, at 3 P.M. Of course, the exact times will vary somewhat with the advance of the season.

If we now decide to look at the situation from the side of the flowers, we can ask ourselves a whole series of questions, such as: What causes certain plants to flower in the spring and others in the summer? What factors regulate the opening and closing of the flowers, once they are there? Are there built-in time clocks such as we found in the bees? Can animal visitors trigger certain events? We shall presently see that different answers must often be given for different flowers. Although at first this may seem a little discouraging or at least confusing, it should not depress us. It merely demonstrates that there is a vast field left for observation where anyone can still make important contributions.

As to the first question (what causes flowers to bloom in a certain season), it is the fashion, nowadays, to distinguish short-day plants, long-day plants, and the so-called day-neutrals. We have to be very careful in applying these names. It is certainly true that short-day plants such as dahlias and chrysanthemums bloom when the days are short, in the fall. What counts, however, is not the actual daylength but the long period of darkness. In some cases it has been shown that if this period is interrupted halfway by a very short period of light, say one of about a minute, blooming is suppressed. Long-night plants would, therefore, be a better name, for there has to be a dark period which exceeds a certain "critical length." On the other hand, in long-day plants such as spinach and henbane which, as we all know, flower in the middle of the summer, a light period of a minimum duration is required. Some of these plants would flower equally well in continuous light. Day-neutrals do not seem to care one way or the other. Many tropical plants are in this group, such as tomatoes, red peppers, and cucumbers.

I wish I could give a real explanation of what, for example, happens in a short-day plant in which flowering is induced by keeping the plant on a long-night regime for several days or weeks. For a while, it was believed the long nights were necessary for the synthesis of a certain hormone which would be directly responsible for the flowering, and this hormone has even been given a name, florigen. In reality, the situation is much more complicated and, to be quite frank, is not well understood at all. Also, light is not always the only factor that should be taken into account. The most striking example of temperature influence that comes to my mind is that of the pigeon orchid, *Dendrobium crumenatum*, of my boyhood days in the tropics. It was quite common in towns like Buitenzorg, where you could find it growing on the bark of trees in many a yard. The delightfully fragrant, pigeon-shaped flowers, snowy-white with a little bit of yellow, would all appear on the same day and in great profusion, giving a magic beauty to the place. Unfortunately, the charm was of a fleeting nature, since the blossoms would not last for more than a single day. A strange thing about this flowering was that one would seldom find pigeon orchids in bloom over large areas. If flowers were out in, say, Buitenzorg, it might well be that none were to be seen in the next town, Depok. Also, their appearance was unpredictable, so that sometimes a blooming period would follow the previous one after seven months, sometimes after five or ten. In European greenhouses, the pigeon orchid *plants* would thrive, but would never bloom at all.

What a mess of strange facts to sort out! People were really at a loss to find an explanation for the strange blooming habits of *D. crumenatum*. The answer to the riddle was finally discovered by an amateur. It is as follows: We like to think of the tropics as an environment where conditions such as temperature and daylength do not fluctuate very much, but this is mostly our imagination. After a heavy tropical rainstorm, sudden temperature drops of 10° to 15° F occur, at least in limited areas, and it is these changes that give the stimulus to the hidden "flower primordia" of the pigeon orchid, which are already present, to grow out and develop into blooms. The growth process itself requires about a week.

This means that, nowadays, we are able to predict the arrival of flowers with a great deal of accuracy, and we can even induce flower appearance in greenhouses, where conditions originally were much too uniform, by giving *Dendrobium* a cold shock.

We have to concede that cases like that of the pigeon orchid are rare. By and large, the temperature in humid tropical regions is indeed rather constant, and for most plants it would be foolish to rely on temperature as the sole factor in the regulation of such vital processes as flower production.

When we consider the actual opening and closing of flowers that are already there, we again find quite a variety of situations. Dandelions, daisies, and some members of the cactus family open in the morning and close in the evening, with the regularity of a clock. Most probably, they have a built-in clock mechanism, for just as in the case of the bees, we find that temperature changes have very little to do with the rhythm. It remains the same even when the plants are kept at a constant temperature. The rhythm will also continue in complete darkness and in constant light, so that the flowers still open at daybreak and close late in the afternoon. It is only very gradually that the opening and closing movements will become less pronounced; the plants eventually stiffen up, so to speak.

However, other flowers are a little more flexible in their behavior. Van der Pijl, who has done excellent work on so-called "clock flowers," has in several cases demonstrated that the hour at which twilight comes is decisive. In *Sida acuta*, for instance, a small relative of our cotton plant, opening of the flower occurs 13½ hours after the twilight hour of the previous day, that is, at about 9 o'clock in the morning. The well-known *Mirabilis jalapa*, which has been used so fruitfully in genetical research, is in the East Indies usually referred to as "pukul ampat" or 4-o'clock plant, because the flowers normally open there a little before 5 o'clock. In Europe, however, they open later in the evening. We must conclude that in this case, too, the moment of opening is determined by the moment of nightfall the day before.

In other instances it is the change from night to day that acts as a trigger. In Chapter 13, "Ambushes and Traps," we shall see how important this is in the inflorescences of certain tropical arum lilies which produce a lot of stench when they open.

As we might expect, some spring flowers are very appreciative of a rise in temperature, to which they respond by unfolding. Among the nicest examples are crocuses and tulips. What happens in a crocus flower when we raise the temperature from, say, 50° to 70° F, is that the upper side of the petals begins to grow much faster than the underside. After a while, however, the upper side ceases growing, while the underside slowly continues to grow. This, of course, means that the underside "catches up" a little. We do indeed observe that the flowers always begin by opening wide, to fold up just a little afterward.

Then, we have those flowers about which we do not really know which is more important, light or temperature. The golden California poppies provide excellent material for study. Even when the cut flowers are kept in water, they will open in the morning and close at night for several days, before the petals finally drop off. Because of this habit of "going to sleep" every evening, and the remarkable pointed-hat shape of the folded corolla, these flowers are often called "nightcaps" in Holland, where they are widely grown as garden plants. In nature, the flowers open when the sun's rays hit them early in the morning. The peaked yellowish cap which enclosed the bud originally and which is formed by the sepals or calyx-lobes is now pushed off, falling to the ground. This habit of casting off the calyx, by the way, is widespread among members of the poppy family. If you are an early riser, you can see how our red garden poppies do it in quite a dramatic way, with an audible noise. When the weather is chilly and foggy, California poppies may be very slow in opening and sometimes do not unfold at all. The petals then just remain tightly closed all day long. Once open in bright weather, the flowers will still respond very nicely by folding up when a big cloud starts covering the sun. Even on a beautiful day, they will begin to close when the shadows begin to lengthen in the afternoon. Long before sunset, we find the petals rolled together again, folded around each other just as in the bud, although not quite so tightly.

Another flower that is extremely sensitive to atmospheric changes is the scarlet pimpernel, *Anagallis arvensis*. It is a small weed of the primrose family which is very common along roads in Europe and the United States. At least the name of the flower is well known because of Baroness Orczy's book and several movies.

However, when one starts looking for it, it is to be remembered that there is a blue-flowered form of this plant which in many places is just as common as the normal red-flowered type. Country-folk call the pimpernel "poor-man's-weatherglass" because the flowers close or fail to open in dull weather; on bright days, they usually fold up around four o'clock.

After having at least mentioned some of the factors that govern the appearance of flowers and their opening and closing, it is now time to pay some attention to the events *within the flowers,* that is, to the appearance of pollen at such a time that self-pollination is prevented. There are countless flowers which know the trick. In some cases the stigmas are only receptive to pollen before the anthers in the same flower are ripe, and in others the pollen is shed first and the stigmas mature afterward.

The name for the phenomenon is dichogamy. In the first chapter we have already seen that it was discovered by Sprengel in fireweed. A whole chapter could easily be devoted to it alone, but here we shall present only one or two examples. Red valerian or Jupiter's beard, a common garden plant, shows the phenomenon very nicely, as illustrated by Figure 22. A case that has great practical importance in America, and which happens to be the most complicated one at the same time, is that of the avocado. During the years 1921–1926, several botanists studied pollination in the avocado very carefully and came up with the following story. In the tropics and in the warmer parts of the United States, there are under cultivation about one hundred varieties of this important food plant, which is pollinated by bees. The varieties fall into two groups, the Class A and Class B varieties. In the former, the flowers have receptive stigmas in the morning and produce pollen in the afternoon, or in other words, they act as females in the morning and as males afterward. In flowers of the Class B variety, the situation is just the reverse, so that they shed pollen in the morning and have receptive stigmas in the afternoon. It is clear that, in the morning, the stigmas of Class A flowers can only be pollinated by insects carrying pollen from Class B flowers; in the afternoon, the stigmas of Class B varieties will be pollinated by pollen from anthers of Class A varieties. It is, therefore, impera-tive that a grove of avocado trees contain varieties of both Class A and Class B, to guarantee cross-pollination. This is especially true for avocado groves in Florida; in other regions, irregularities in the timing of the dichogamy occur, so that interplanting is unnecessary, and even the trees of a single variety may set fruit. Bees, however, are still necessary because the pollen is too heavy and sticky to be dispersed by the wind.

As to our fourth question, whether animal visitors can trigger certain timing mechanisms in a flower, there are indeed plenty of cases where the answer must be Yes. Barberry is a case in point. Looking down upon the beautiful yellow flowers from above, we

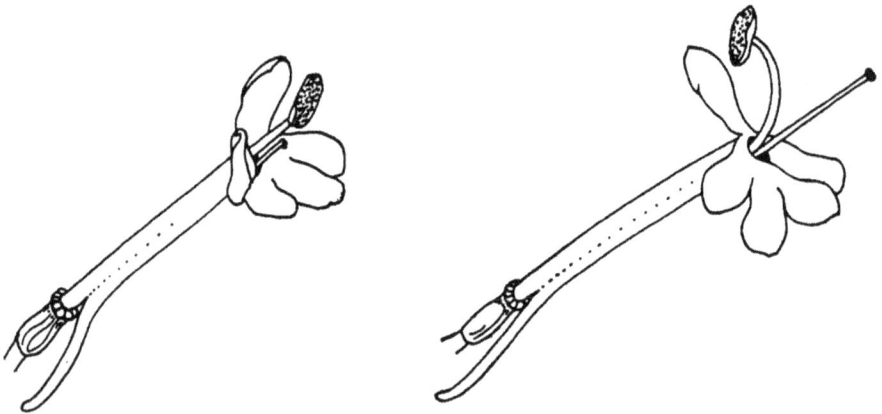

Figure 22. Two flowering stages in red valerian (*Centranthus ruber*). The stamen, which ripens before the stigma, moves out of the way when the latter reaches maturity. This is intended to prevent self-pollination.

notice that the six "petals" are hollow, each one sheltering a stamen (Fig. 23). The T-shape of the ripe stamens is caused by the fact that each anther has, on its right and left side, a little "lid" that opens wide to release the pollen. For the time being, however, most of the pollen remains attached to these anther lids. Now, looking at the base of each stamen, we see that there is a dark-orange nectar gland on each side of it, borne by the petal. The nectar that is secreted will therefore accumulate in the depth of the flower around the base of the stamens. A bee sticking its proboscis into the flower to get at this nectar just cannot avoid touching the lower part of the filament. This, however, happens to be very sensitive, and the stamens will therefore immediately begin to curve inward, so that the pollen-covered anther lids will rapidly swing toward the center of the flower, powdering the bee's proboscis. There is a good chance that the bee will leave the pollen

Figure 23. Moving stamens in the flowers of barberry (*Berberis*). Top of picture: flower-bearing twig. Lower right: open flower showing arrangement of nectaries and T-shaped stamens. Lower left: stamen moves inward toward pistil when its base is stimulated by a needle. In nature, the stimulus is provided by the proboscis of a visiting bee trying to reach the nectar.

which it has thus received on the stigma of the next flower. The anther that was stimulated will, after a short while, return to its original position in the hollow of the petal, so that it is ready for the next visitor. It is the easiest thing in the world for us to imitate the activity of the bee with a needle.

If no barberry flowers are handy, we can try the stamens of *Sparmannia,* a plant which many people keep in their homes for decorative purposes. Many cactuses, too, have very sensitive stamens, a matter to which we will return in Chapter 17, "On the Wings of the Night."

As to sensitive stigmas, a beautiful example is provided by the various monkey-flowers (*Mimulus*) that are so abundant in wet places in the United States. Our photograph (Fig. 24) shows that in a fresh flower of *Mimulus Lewisii* the two broad, roundish stigma lobes are spread far apart. Suppose, now, that the probing tongue (proboscis) of a bee, covered with pollen grains from a previous visit to a monkey-flower, tries to penetrate the narrow corolla tube. In its path it will find the lower lip of the stigma, which will act like a shovel and literally scrape the pollen from the upper side of the proboscis. As a result of the contact, the two stigma lobes will immediately begin to come together, like the leaves of a book. The pollen will thus be caught very firmly and will find itself in a little moist chamber where the chances for its germination are ideal. Meanwhile, the bee's proboscis, now clean, will continue inward and will touch the anthers; these will open and release their pollen, so that the proboscis will again be loaded. When the bee finally withdraws its mouth parts, there is no chance that the precious dust will go to the stigma, since that organ is still folded up and no longer stands in the way. If the next monkey-flower visited is a fresh one, the pollen will be scooped up again by the stigmatic shovel. Our *Mimulus* flowers, then, kill two birds with one stone. It is just as if they want to make sure that self-pollination does not take place and that the pollen from another *Mimulus* flower gets an excellent chance. Yet there is even more to the events in *Mimulus* than meets the eye. It turns out that one can imitate the movement of the bee's tongue with a needle, forcing the stigma lobes to come together. In this case, however, they will very soon come apart again—*which they will not do when they receive pollen from the same species.* Pollen from a different species does not cause permanent closing of the stigma lobes either, so that the latter soon occupy their original position again, as if waiting to receive the "right" pollen. Undoubtedly, the pollen of the same species, which germinates in the little moist chamber formed by the stigma lobes, produces some chemical substance which induces

Figure 24. Sensitive stigma lobes in one of the monkey-flowers, *Mimulus Lewisii*. Left: before touching the lobes. Right: afterward.

the flower to keep these lobes together. What a magnificent arrangement, and what a fertile field for future studies! And even this may not be all. Van der Pijl believes that in *Mimulus* flowers pollinated by hummingbirds the closed stigma lobes act as a signal, telling the bird to avoid this flower that has just given up its nectar. We hasten to add, however, that this idea was not borne out by recent observations.

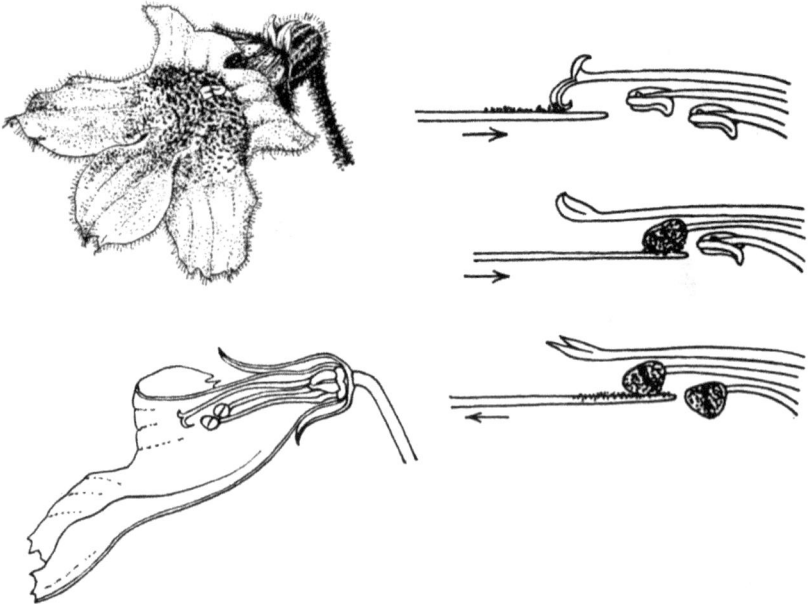

Figure 25. Sensitive stigma lobes in *Rehmannia,* a relative of the monkey-flower, *Mimulus.* When the lower of the two lobes is touched by a bee's proboscis in the manner indicated, the pollen on that proboscis is scraped off for safekeeping in the tiny, moist chamber formed when the two lobes meet. Subsequently, the bee withdraws pollen without contaminating the stigma of the same flower.

The flowers of bladderwort (*Utricularia*) which one can sometimes find in the very lakes and ponds that are flanked by *Mimulus,* show the same behavior of the stigma lobes. Good places to find them are the swampy regions of, say, New Jersey. I remember having seen magnificent big bladderwort flowers growing out of the water there. In a region where both monkey-flower and bladderwort are rare, the flowers of the *Catalpa* tree might be tried. In many parts of the United States, this tree is quite common in parks and on university campuses. To those fortunate enough to have a greenhouse, we recommend the beautiful flowers of *Rehmannia* (Fig. 25) and *Torenia*, relatives of the monkey-flower and the *Catalpa* tree. What differs in the different cases is the length of time the stigma lobes remain folded together when the experiment with the needle is performed. With the exception of *Mimulus*, I do not believe that it is ever more than 2 minutes—but this can hardly be called a drawback.

# 9

# A Drink, Some Snuff, and a Whiff of Perfume

## NECTAR AND NECTARIES

"The sweet sap secreted in the depth of the flowers and collected zealously by the bees and other insects, probably is a true honey, albeit one that is still very thin and fluid." Thus spoke Joseph Gottlieb Koelreuter in a book published between 1761 and 1766. Since he also recognized the importance of cross-pollination (although he underestimated it; see pages 9–10), it is fair to say that he was the first to appreciate, from more than one angle, the true "meaning" of nectar. We have also seen how the presence of nectar drops in the flowers of a wild geranium gave Sprengel his first impetus, leading finally to the establishment of his marvelous and fairly complete cross-pollination theory. And yet, really to become fully and completely aware of the importance of nectar in pollination, it is necessary for us to follow in our minds the development of flowers through the ages.

What type of flower occurred on the ancestors of our present-day flowering plants, millions of years ago—let us say in the Mesozoic era before the Cretaceous? As a matter of fact, this is not a question that is well put. Obviously, the unknown predecessors themselves were not flowering plants yet, so it is impossible that they possessed "flowers" in the modern sense of the word. Perhaps it would be safer to say that they had "flowering structures," and the chances are that these did not possess both stamens and pistils, but that there were separate male, or stamen-bearing structures, and female ones bearing only pistils.

The pollen produced by the stamens was blown away by the wind, and pollination depended largely on chance. Much more pollen had to be produced than was really needed to fertilize the ovules. Nowadays, we still find such a situation in evergreens such as fir, pine, and hemlock, and since we know that the pollen here is almost odorless, we assume that in the primitive ancestors of the flowering plants it was likewise without scent.

Primitive insects, therefore, cannot have been very much interested in this pollen. They did not have much of a chance to get at it, anyway, because it was blown away as soon as it fell out of the pollen sacs. In her efforts to improve upon the existing situation and to avoid waste, Nature then must have begun to make some pollen grains a little sticky, so that they would tend to clump together and would not fall out of the pollen sacs so easily. Insects now had a chance to eat the pollen. There was also the possibility that some of the grains would remain intact and would stick to the bodies of the insects, which could now transport them to the pistils. But another difficulty had to be conquered at the same time. A pollen-eating insect must, of necessity, have biting mouth parts. And the chance that such an insect would be interested in the pistils is very slim indeed. In those primitive forms which are still in existence, the pistil often produces an attractive sweet fluid which can be lapped up but not chewed on, and we have no reason to think that it was different for the extinct forms.

It can be readily seen that there are at least two solutions to this problem. In the first place, Nature has combined the male and female "flowers" so that a flower of the modern type, a so-called "hermaphroditic flower," with both stamens and pistils, was obtained. One and the same insect could now find something to its liking, no matter what flower it chose to visit. Insects which had only biting mouth parts could be just as useful as those with sucking organs. This arrangement has, of course, led to the development of insect types that can take advantage of both pollen and sweet fluids. The other solution was to leave the male and female flowers separate, but to provide both with a sweet nectar. How effective such an arrangement is can be seen very nicely in the spring when the pussy willows are out. Both the male and the female catkins attract quite a few bees and bumblebees, and when one examines the individual little flowers very carefully, it will be seen that they all have a nectar gland at their base (Fig. 26). Perhaps this example will cause a raising of eyebrows, and indeed it may be somewhat unfair to use it. We know nowadays that the not too remote ancestors of the pussy willows had hermaphroditic flowers, and that their descendants have simply gone back to the

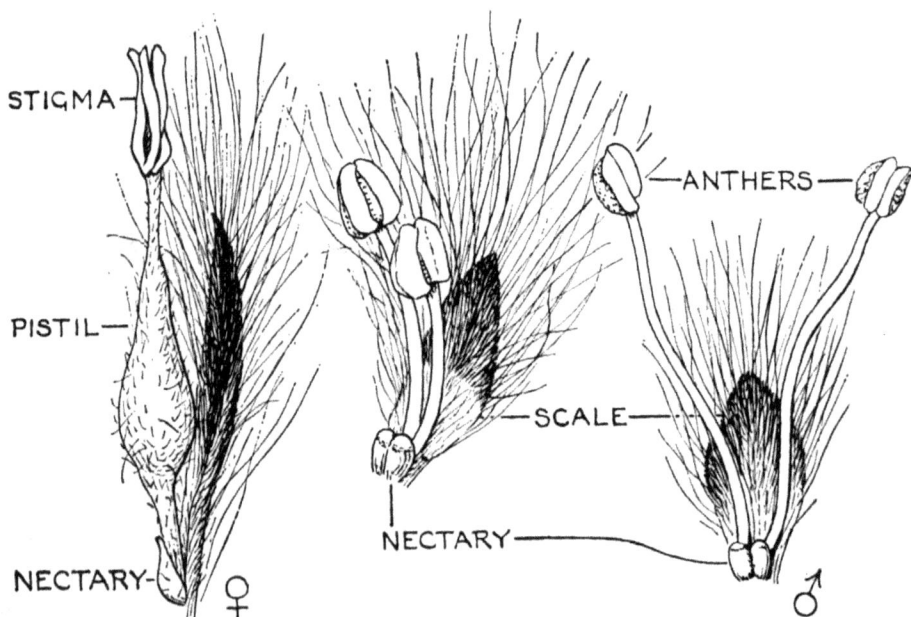

Figure 26. A female flower and two male flowers of a willow, showing the nectaries.

more primitive condition. Nevertheless, they do illustrate the point.

Another question that could be asked is how Nature could so suddenly "conjure up" these nectar-producing glands. Did they really appear like a bolt from the blue? The answer is that they did not have to be conjured up because they had been available all the time. In order to explain this rather amazing situation, it should be pointed out that, as a matter of fact, a flower is only a shoot which has undergone some sort of metamorphosis. Imagine an "axis" on which there are placed a number of leaf whorls. The axis is made shorter, so that the leaves are brought close together; the leaves in the lower whorls are given an attractive color and, finally, the shape of the upper ones is changed so that they now have the appearance of stamens and pistils. It is quite likely that most people would agree to call this a flower. The main thing to remember is that stamens and pistils as well as the calyx lobes and the petals in the corolla are leaves, or sometimes leaves fused together. Now it so happens that we are familiar with many cases in which leaves (ordinary leaves) have nectar glands. Cherry leaves are a fine example. Often we find a couple of small, wartlike glands at their base, one on each side of the leaf stalk. It is not at all amazing that they can produce sweet sap, for we know that, in the light, leaves will carry out that phenomenal process of photosynthesis in which they build up an abundance of sugars (and other substances) from the carbon dioxide which they find in the air. It is really tempting to dwell on these *extra-floral nectaries,* as they are called; often they are the centers of quite a little animal community, and ants especially are frequent visitors. But we have to talk about flowers. I hope that, after what has just been said, the presence of nectar glands in flowers will no longer cause amazement.

In modern flowers, nectar can be produced in all kinds of places. In many of the monocotyledonous plants (lilies, irises, and their relatives), the base of the pistil is the active spot. The nectar that accumulates may eventually fill up the whole flower tube. In the flowers of the Umbelliferae and in those of English ivy, there is a disk of special nectar-producing tissue surounding the pistil (Fig. 27). Since the flowers here are of a very open type, the nectar is fully exposed to sun and dry air and will as a result become very thick—so much so, in fact, that it can only be utilized by insects such as flies, who are able to redissolve it with their own saliva. In most cases, though—and Sprengel knew this—the nectar is protected from the atmosphere, either by means of hairs (see page 11) or by the fact that it is hidden in the flower tube or in a special spur. Good examples of spurs are found in nasturtium, columbine, and red valerian and also in some of our violets and terrestrial orchids (Figs. 22, 28, 29, and 30; also see Fig. 68, page 176).

Quite often, certain elements ("leaves") that helped to form the original flower have undergone a metamorphosis to become special "nectar leaves." (It should be added that, as a rule, they are also largely responsible for the characteristic flower smell.)

These particular nectaries vary widely in size and shape. However, they are often somewhat club-shaped or have the appearance of an open bag sitting on a small stem, which has led to the belief that in this case they are derived from stamens. A number of examples are to be found in Figure 28. In the buttercups, the nectar leaves form at the same time the showy part of the flower, attracting the insects. Most people simply refer to them as petals. The flowers

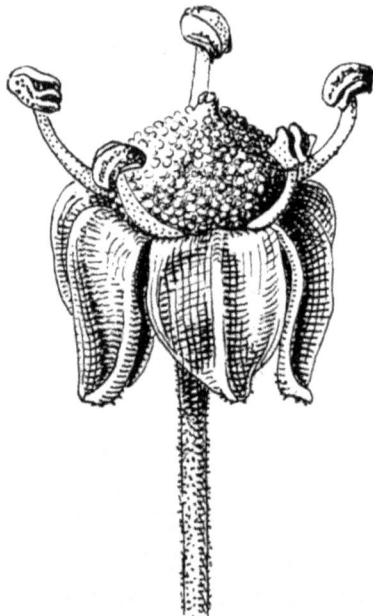

Figure 27. Flower of English ivy *(Hedera helix)*, with numerous nectar droplets on the base of the pistil (7 ×).

of *Ranunculus acer*, one of our commonest buttercups, have a tiny scale at the base of each "petal," under which the nectar is hidden (Fig. 31). Still somewhat enigmatic are the nectar leaves in *Parnassia* flowers—five hand-shaped, fragrant structures, each having two small nectar glands on its upper surface. Each of the many "fingers" is tipped by a little ball (Fig. 32) which glitters and shimmers in the sunshine like a real drop of nectar, although no fluid is present there at all. These are the famous "false nectaries" of *Parnassia,* the purpose of which, according to Sprengel, is to fool the "dumb flies," attracting them from a short distance and finally leading them to the true nectar glands. Although this idea, like so many others advanced by Sprengel, has often been pooh-poohed, modern experiments with the fly *Lucilia* lead us to believe that it may very well be correct.

In my student days, I was fortunate enough to have the opportunity of analyzing the nectar from some 15 different flowers with the aid of a clever fermentation method invented by my old professor. Although there was a great deal of variation among the different species, I found that the nectar from one and the same species is often amazingly constant. My champ was horse chestnut, with approximately 70 per cent sugar. (Ruth Beutler found 66 per cent here, which is quite a satisfactory agreement.) Common rue is a good second, with 66 per cent, and *Asclepias cornuti,* one of the milkweeds, has almost 60 per cent. *Tilia euchlora,* a species of linden, boasts 65 per cent; *T. vulgaris* only 50 per cent—on the average, for in this case the percentage of sugar varies somewhat with the weather. In the literature, I found only a few plants with a sugar concentration appreciably higher than horse chestnut. According to Elisabeth Kleber, marjoram, *Origanum vulgare,* has 76 per cent sugar, while according to Fahn the blossoms of apple (*Pyrus Malus*) can produce a nectar with 87 per cent sugar and those of *Satureja* one with 85 per cent. A good average for all the different flowers visited by bees might be 30 to 35 per cent. In this, I am in agreement with Ruth Beutler again.

It stands to reason that the figure for the *sugar concentration* is not all-conclusive. The *total quantity* of nectar secreted each day should be taken into account, too. In this respect, there are large differences. A very good producer among the bee plants is *A. cornuti,* with 5.3 mg per day per flower. In this connection, I might mention that certain American Indians sometimes collected milkweed nectar for human consumption. *Tilia vulgaris* produces only 1.9 mg per day. Bird-pollinated flowers usually have a great abundance of nectar which, however, is rather dilute (see Chapter 10, "Something for the Birds").

As to the nature of the nectar sugars, the bulk of it probably is sucrose, the same sweetening we use for human consumption under the name of table sugar. The reason it is a little difficult to give ironclad data about the exact sugar composition of nectar is that it is very, very easy for the sucrose molecule to react with water and to fall apart, which gives us the two building blocks of sucrose, namely, glucose (or grape sugar) and fructose. This happens, for

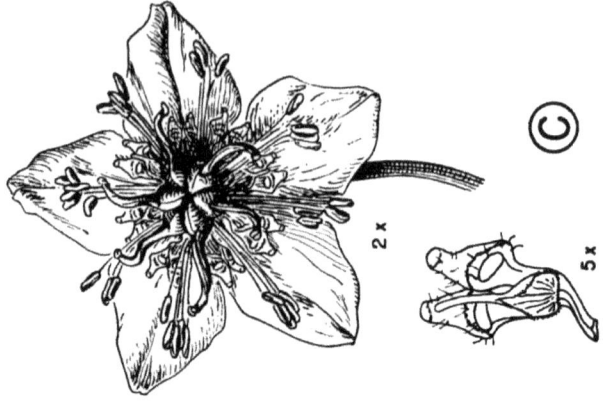

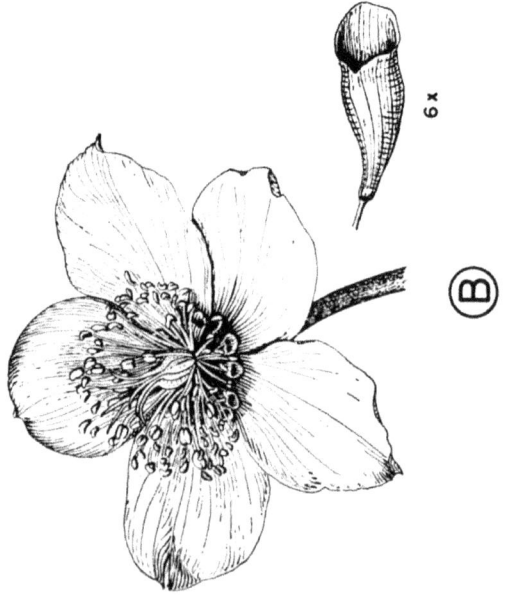

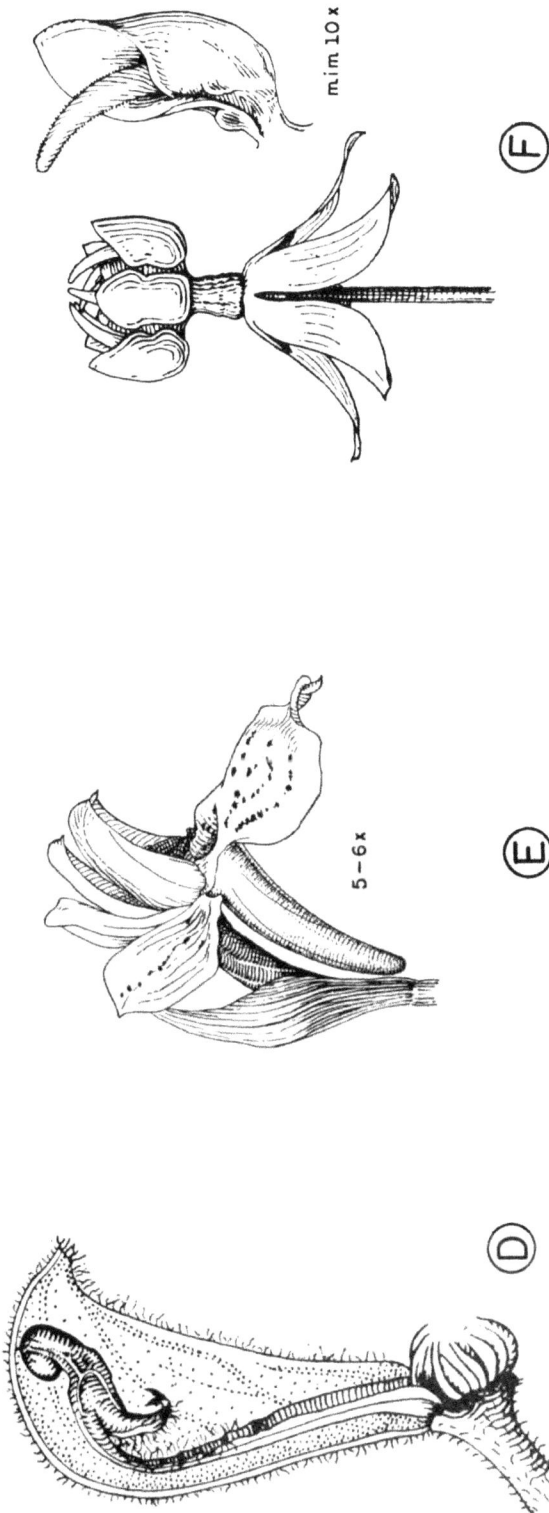

Figure 28. Various types of nectaries. (A) Winter aconite (*Eranthis hiemalis*) and (B) Christmas rose (*Helleborus niger*), both with easily accessible nectar in pitcher-shaped containers, (C) love-in-a-mist (*Nigella*), (D) monkshood (*Aconitum*), and (E) milkweed (*Asclepias*). The complicated nectaries of love-in-a-mist, resembling miniature flowers, can only be handled properly by bees. Monkshood is considered to be a typical bumblebee-flower. (F) Flower of *Orchis maculata*, a European terrestrial orchid in which the nectar is secreted into a long spur.

instance, when nectar is contaminated with certain yeasts, a thing which is not at all uncommon in nature. There is a whole group of yeasts that thrive in nectar, and ants sometimes carry these from one flower to another on their legs. To the bees, it does not make too much difference whether the original sucrose is present

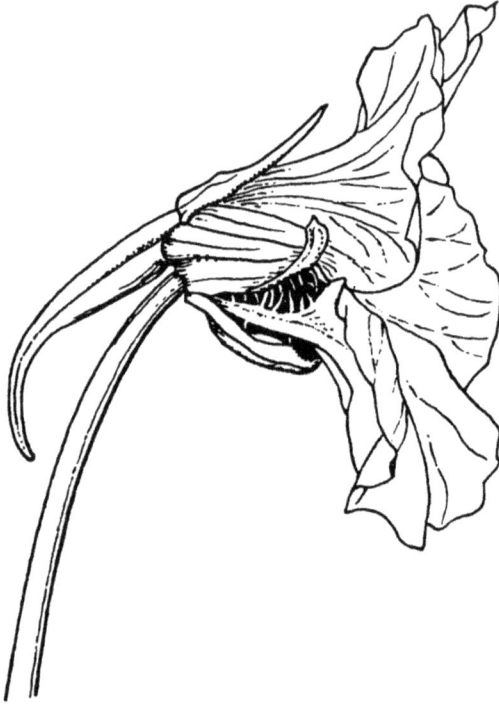

Figure 29. Flower of *Nasturtium,* showing the spur.

or a mixture of equal parts of glucose and fructose, the so-called invert sugar. As we shall presently see, all three sugars taste sweet to the animals. To be sure, some investigators claim that bees find the invert sugar a little sweeter than the original sucrose, but the difference cannot be great. Ultimately, all the sucrose will be converted to glucose and fructose by the bees, anyway, with the aid of an enzyme in their bodies called "invertase." This process is somewhat comparable to the conversion of starch to sugar in our own bodies, with the aid of the "amylase" present in our saliva. In contrast to invert sugar, sucrose crystallizes easily. Since honey is a material in which crystallization is undesirable, the elimination of sucrose from the nectar must be of great advantage to the bees.

Figure 30. Flower of columbine (*Aquilegia*) displaying five nectar-containing spurs.

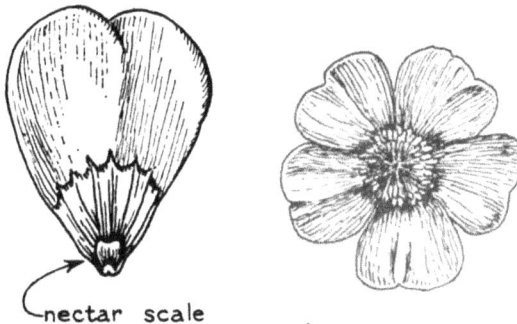

nectar scale

Figure 31. Buttercup flower and separate buttercup petal showing the "nectar scale."

It was again von Frisch who, in 1919, began an investigation of
the sense of taste in bees which is, as one would expect, located in
the mouth parts. It turns out that there are quite a few substances
that seem sweet to us but are not recognized as such by bees. All
in all, von Frisch has tested 34 sugars and sugar derivatives which

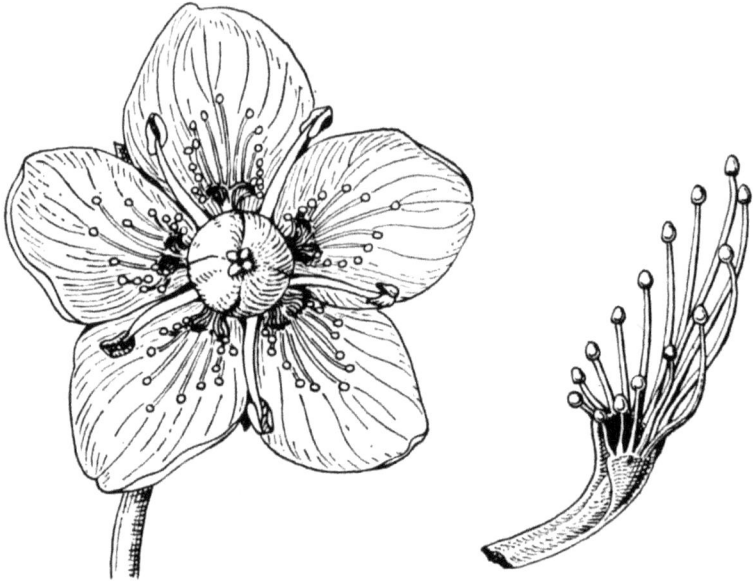

Figure 32. False nectaries in the flowers of grass-of-Parnassus (*Parnassia palustris*).

to us are decidedly sweet. Only 9 of these were also sweet to the
bees, namely, sucrose or table sugar, maltose or malt sugar, fructose,
glucose or grape sugar, and 5 lesser-known compounds. But even
in their sensitivity to a sugar such as sucrose, man and bee differ.
How do we find this sensitivity? It is not enough to just offer the
bees various sugar solutions. If we offer a 20 per cent sucrose, the
bees will suck it up. A 10 per cent solution is accepted by some,
refused by others. A 5 per cent solution is tasted, but refused by
all. But this gives us only the *threshold of acceptance*. It is not
even a constant value because, if there are many flowers with a
concentrated nectar around, such as horse chestnut, the threshold
may be as high as 40 per cent. In order to find the *threshold of perception*, we have to starve the bees first, and then give them the
choice between water and sugar solutions of different concentrations. The value found this time is 1 to 2 per cent sucrose, equiv-

alent with 10 to 20 grams per liter. Lower concentrations are confused with water, which means that the bees do not taste these
as sweet any more. It is a good thing that bees normally do not
accept solutions of low concentration, for these would never yield
a honey that would keep in the hive through the winter months.

To see things in their proper perspective, von Frisch compared
the sense of taste in the honeybee, man, minnow (which can taste
with its whole skin), and the red admiral butterfly, which (as we
shall see in Chapter 16) has taste organs in its feet. In order to be
tasted as sweet, a sugar solution must have per liter:

28.4   grams of sucrose for a bee
4.25   grams for man
0.068  grams for a minnow
0.027  grams for the red admiral butterfly

We notice that the bee does not even come out second best. The
taste organs of the legs of butterflies are the most sensitive ones
known to date. On the basis of the above figures, we find that those
of the red admiral are 150 times as sensitive as the human tongue!

By means of very ingenious experiments upon which, unfortunately, we cannot dwell here, von Frisch could show that bees, like
man, can distinguish four taste qualities: sweet, bitter, sour, and
salt. They are more sensitive to sour and salty substances than we
are, but are more tolerant of bitter tastes. It is therefore possible
to add to sucrose substances which are very bitter from the human
point of view, yet do not make the sugar unacceptable to the bees.
This is the basis of the use in many European countries of octo-
acetyl sucrose or octosan. The substance is simply added to sucrose,
which as a result becomes completely unfit for human consumption
and can therefore be sold, cheap and without any risk of fraud, to
the beekeepers to feed their animals when these are having a hard
time.

## POLLEN

In spite of the tremendous importance of nectar, which we have
just emphasized, there are many insect-pollinated flowers in this
world which do not contain any nectar whatsoever. They compen-
sate for the lack by producing huge quantities of a sticky pollen.
This type of "pollen flower" can immediately be distinguished from

the more primitive types such as we find in pine, fir, and other conifers, because its color is conspicuous and not green and, moreover,
because there is a very noticeable smell. The main pollinators are
various types of bees, which use the pollen grains for their beebread,
and certain beetles which devour large quantities of pollen on the
spot. Needless to say, then, these flowers produce many, many more
pollen grains than are actually needed for pollination itself.
Furthermore, the flowers are of a simple and open type, offering
their wares freely. Again, this is only what we can expect, for
beetles are by no means so adaptable as, for example, hummingbirds, and a complicated flower would be "over their heads."

After this introduction, you will probably have no difficulty in
recognizing wild roses, peonies, and red poppies as members of this
group. It is also safe to include the buttercups, although (as we
have just seen) these do produce a little bit of nectar under the
minute scale which they have on each petal.

From the point of view of the flowers, of course, it is a deplorable
thing that so much good, viable pollen disappears into the hungry
mouths of the visitors. Some plants, such as *Cassia* in the pea
family, have found a nice compromise by producing in one and the
same flower two types of anthers, normal ones which produce
healthy pollen and others which yield sterile but very tasty cells
constituting an excellent food for the visitors. While robbing these
"food anthers" of their contents, the insects bring about pollination,
or at least they will be covered with viable pollen that can be used
in another flower.

Some orchids have hit upon still another solution. In 1886,
J. M. Janse found that the flowers of a tropical orchid (a *Maxillaria*) have on their lower lip a group of hairs which easily fall apart
into individual cells. In his case, these cells were rich in starch.
O. Porsch later described similar *"food hairs"* with fat and protein,
for example, in the Brazilian *Maxillaria, M. rufescens*. Such nutritive tissues in orchids form a good substitute for nectar and for the
pollen which, in this case, cannot be offered to the insects for food
because, as we shall presently see, it comes in neat little packages,
pollinia, and sacrificing a few grains would sacrifice the whole
pollinium. In some of the native orchids of western Europe and
America (*Orchis* species), there is not too much difference between

the nutritive tissue that forms the lining of the spur and true nectar. The cells are very rich in sugar and will yield a sweet syrup as soon as they are touched by the probing proboscis of the pollinating hawkmoth.

In some of the most highly developed plants that are insect-pollinated, we find that the pollen is liberated and transported in groups of four, the so-called "tetrads." (It so happens that pollen grains are always *formed* in groups of four, but it would take us too far afield to explain this.) In still other flowers, such as those of evening-primrose, fuchsia, and fireweed, we see that larger groups of pollen grains are connected by sticky threads or even nets of a gluey material that is usually referred to as *viscin* because it is so similar to the birdlime prepared from the berries of the mistletoe, *Viscum* (see Fig. 33). In the plants of the milkweed family (Asclepiadaceae) and in most of the orchids, the association of pollen grains is even closer. All the grains from one pollen sac or from one anther form a compact but not sticky mass, a pollinium (Fig. 34). This is always connected with a sticky disk or a clasplike "translator," the idea being that these latter structures will attach themselves to the proboscis or the leg of a visiting insect, so that when it flies away it will carry the pollinium with it (Fig. 35). In Chapter 13, "Ambushes and Traps," we shall explain more fully how this happens in milkweed flowers. We shall also have the opportunity there to discuss the situation in the lady's-slipper orchids (*Cypripedium* species) where the pollen grains form a very sticky, oily mass, not a true pollinium. Right now we will only comment on one point you may be wondering about: how do the pollen grains that together form one pollinium germinate once they are on the right stigma? The answer is that there will emerge from a preconditioned spot on the pollinium (Fig. 34) a whole bundle of pollen tubes which will grow down through the pistil like a thick cord, to reach the ovules. It is not at all difficult to observe the germination in this case (and in many others, for that matter). All we have to do is to put the pollinium in a rather strong sugar solution, let us say 20 to 50 per cent sucrose, in a small watch glass. After a few hours or a day we check the result under the microscope. If we do the experiment with a large enough number of pollens from different species, we will be struck by the fact that each has its distinctive

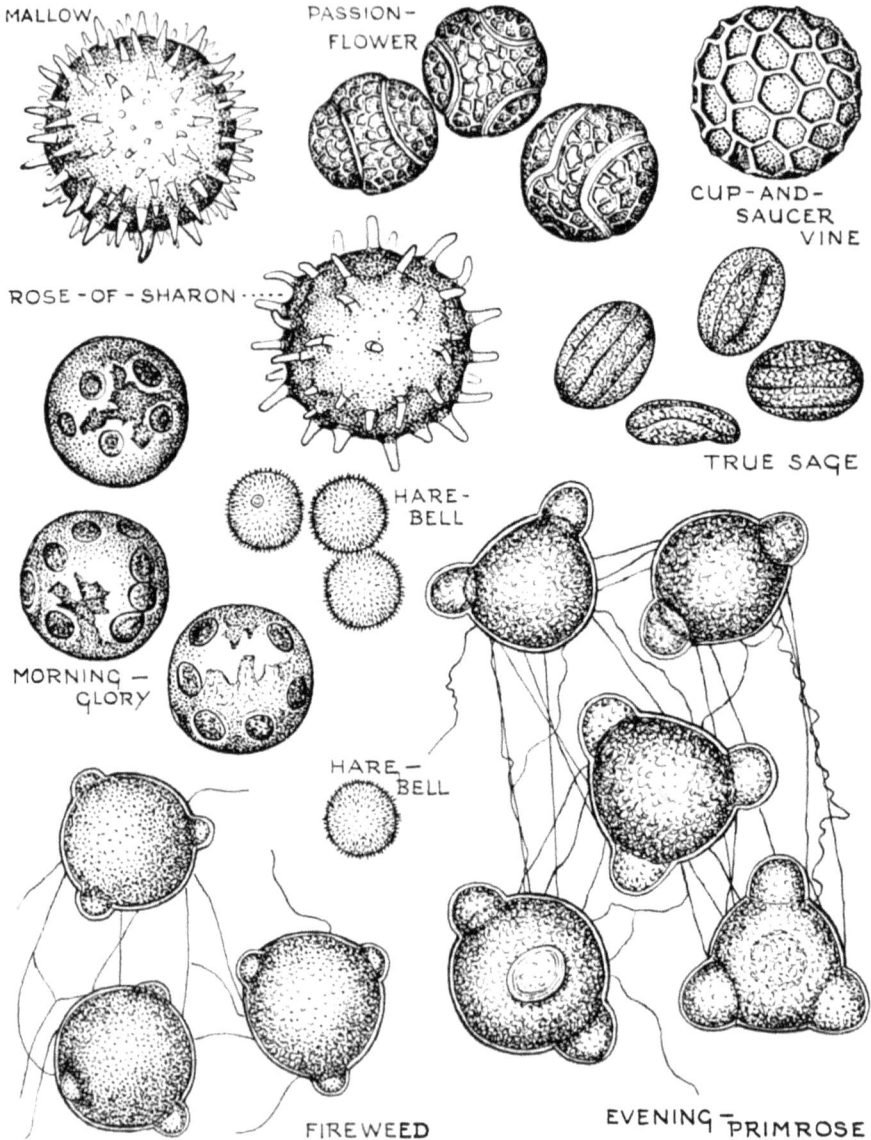

Figure 33. Pollen grains of different shapes and sizes. Upper row, from left to right: mallow (*Malva moschata*), 0.14 mm; passion-flower (*Passiflora*), 0.07–0.075 mm; cup-and-saucer vine (*Cobaea scandens*), 0.10 mm. Second row, from left to right; morning-glory (*Convolvulus sepium*), 0.08 mm; rose-of-Sharon (*Hibiscus syriacus*), 0.15 mm; true sage (*Salvia glutinosa*), 0.06 mm. Lower left: fireweed (*Epilobium angustifolium*), 0.08–0.09 mm. Lower right: evening-primrose (*Oenothera biennis*), 0.10–0.12 mm. The very small, round grains (middle left) belong to harebell, a *Campanula* species, 0.05 mm. Notice the threads of "viscin" connecting the grains in the fireweed and evening-primrose.

shape and size of grain (see Figure 33, left, and Figure 84, page 213). Pollen grains from wind-pollinated plants are, as a rule, roundish and smooth; insect-borne pollens, on the other hand, are rough and irregular or are provided with a tremendous number of

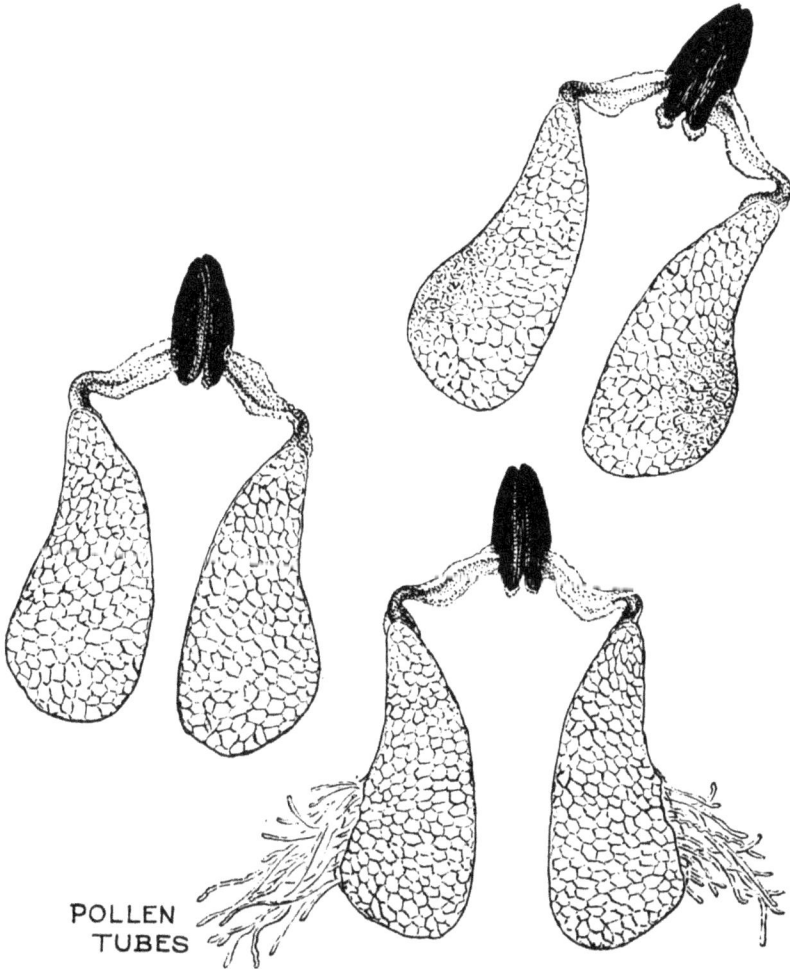

POLLEN
TUBES

Figure 34. Pollinia of milkweed, *Asclepias*. Lower right: after "germination" in sugar water.

tiny teeth (as we see in squash pollen, for instance). Thus, insect pollens can easily cling to the hairy legs or bodies of the transporting animals. Many books state that wind-borne pollen grains are smaller than those of the other type. However, this is not an ironclad rule, for the smallest pollen grains known are those of a little forget-me-not which is insect-pollinated. Stickiness is a much

safer characteristic to use if one wants to distinguish between the
two pollen types. Needless to say, it always occurs in insect-borne
pollens.

Figure 35. Dead honeybee collected on a milkweed flower, with pollinia on
its legs and tongue.

## SMELL

As we mentioned earlier, modern pollen flowers often have a
very pronounced smell, and it is certain that this originates in the
anthers and the pollen grains. Perhaps the adhesive that is so characteristic of insect-borne pollens has something to do with it. As a
working hypothesis we may submit that, as soon as the pollen smell
occurred in the wind-pollinated ancestors of the modern flowering
plants, those stamens that possessed it acquired the magic property
of attracting certain insects from a small distance. We have already
seen that certain nectaries can be thought of as derivatives of
stamens (Fig. 28). It is, therefore, not at all strange to find that

they have a fragrance, too. It is possible that in those flowers where the emphasis was put more and more on the nectar, the production of fragrance was shifted almost entirely from the pollen sacs to the nectaries.

There are indeed a great many cases where the characteristic scent of a flower must be ascribed to the "nectar leaves" and the nectar. [This scent is a very important factor in a phenomenon which we have not yet discussed, the so-called *flower constancy* of bees and probably of other insects as well.] By flower constancy we mean the habit of bees to restrict their visits to one species of flower—at least for a certain length of time. It is obvious that it acts as a terrific timesaving device for the insects and as a boon to the flowers that have to be pollinated, because they are now sure to receive the right pollen.

Of course, it would never do to claim that odor is the only factor in flower constancy. Earlier, in our discussion of color sense in *Bombylius*, we have tried to make it plain that the visit of an insect to a flower is a complicated chain of events. First, the general color of a flower may attract the animal from a distance, but this is soon followed by other stimuli and responses, and such things as honey-guides and smell come into play. Still, it is well to remember that a honeybee can distinguish only four colors, so that the animal would almost certainly be led astray if, in its quest for a given food source, it relied on color sense alone. Furthermore, although form perception does play a role in a number of cases—such as that of bumblebees and hound's-tongue, where, according to A. Manning, the shape of the whole plant is decisive—it can hardly be expected that the shape of a particular flower is very critical. Form perception in bees is simply too poor for that. To a certain extent, then, we are justified in singling out the smell for a special discussion.

Being interested in all the aspects of bee life, von Frisch has, since 1918, carried out a whole series of experiments on their sense of smell. It was not difficult for him to demonstrate that bees can be trained to certain smells, just as they can to colors. In general, substances fragrant to us turned out to be fragrant to bees as well. Conversely, compounds considered odorless by man were, most often, so regarded by bees. However, it was almost impossible for von Frisch to train bees to the smells associated with rotting meat or fish. This ties in nicely with the observation that flowers with a carrion smell, such as those of *Stapelia, Rafflesia, Amorphophallus,* and certain arums, are pollinated by flies and not by bees.

There is not too much difference between man and bee in sensitivity to smells. The bee, however, has the edge on us when it has to pick out a certain smell from a number of others. Its ability, in this respect, is downright uncanny! In a particular experiment, von Frisch compared the essential oil to which his bees had been trained (obtained from the skins of Italian oranges) with 23 other oils offered side by side. Some slight confusion was noticeable only

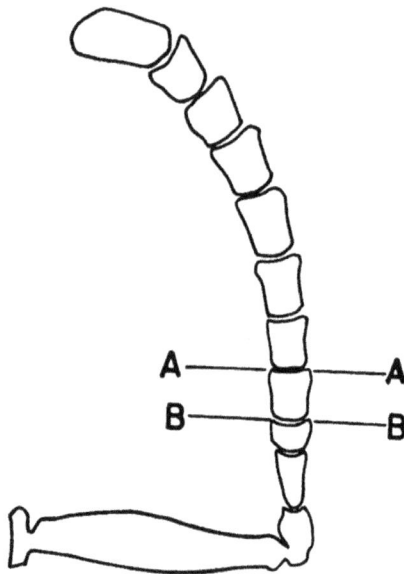

Figure 36. Antenna of a honeybee, somewhat diagrammatical. The organs for the sense of smell are to be found in the eight segments forming the tip. Amputation of seven of these (cut A–A) still leaves the animal with a sense of smell. Removal of all eight (cut B–B) eliminates it completely.

with the 3 essential oils which, like the "training oil," had been prepared from citrus fruits.

The seat of the olfactory sense is in the upper segments of the bee's feelers or antennae (Fig. 36). Each antenna has, on a special base piece, 11 segments covered with the same hard substance, chitin, that protects the whole animal. The first 8 (reckoned from the tip) have pores in the chitin which lead to the various, microscopically small sense organs that can perceive the smell. In the basal 3 segments, these pores and sense organs are lacking. A bee that was trained to a certain smell did not respond to that smell any more when the 16 ($= 2 \times 8$) terminal segments of the antennae were amputated. In various ways, von Frisch was able to show that this result was not due to shock or pain. His amputee bees continued to look for food vigorously, they still responded to the color to which they had been trained (preferring yellow over blue) and they still responded to the training smell if just one of the segments was left intact. The story therefore appears to be foolproof.

One great advantage of having antennae instead of something like the human nose is that antennae can be moved about so much more easily. The insect can, therefore, get some "mental" image of the shape of the odoriferous object. In this connection, it should be kept in mind that the organs for the sense of touch are located in the antennae, too; they and the olfactory organs record their impressions simultaneously. So, a round, scented object may give a bee quite a different sensation from an angular one. Bees may "smell the form of an object," so to speak!

Another result of the mobility of the antennae (combined with their small size) is that scent patterns in flowers that escape us completely can easily be picked up by the bees. That such scent patterns exist is certain. For instance, by collecting the differently colored parts of a narcissus and training bees to their scent, it can be shown that they differ in the quality of their aroma. All this strengthens our belief, hesitantly expressed in Chapter 5, "Signposts of All Sorts," that flowers possess olfactory nectar-guides in addition to the visual ones.

*Part II*

# COMPANIONS
# OF THE WIND

# 10

## Something for the Birds

Not far from the place where Washington crossed the Delaware, there is a beautiful park where specimen plants of all the wild-flowers of Pennsylvania have been assembled. It was here that I saw my first wild hummingbird, a rubythroat. We were watching some magnificent, waist-high plants of the cardinal flower, *Lobelia cardinalis*. In the stillness of the late summer afternoon, they resembled frozen torches, aflame as they were with a dazzling display of scarlet blooms. And then, as a sudden note of contrast, there was the dart of a flying animal, heading straight for the flowers. A hawkmoth, I thought, as any European greenhorn would have done. But suddenly, not more than a yard away from me, the visitor came to a complete standstill in mid-air, the wings just a blur but the head so completely motionless that I could see the twinkle in the animal's eye. And a gay twinkle it was—not at all like the glassy stare of a hawkmoth. Showing no fear at all, the hummingbird began to sip nectar from one *Lobelia* flower after another, hovering, darting, flying backward sometimes, its breast feathers a coppery mail which reflected the sun's lightshafts in ever-changing colors. In short, it was a performance that sent shivers down my spine.

To those of my friends who might want to take these words with a grain of salt, I simply say that you have to be a European to understand my feelings. In all Europe, there are no birds which are interested in nectar or pollen for food; the closest place where one could find them would be Palestine (the area before partition in 1948). As a result, there has always been a tremendous emphasis on insects in European pollination work, and the birds have been downright neglected. In a region such as South America, this would have been unthinkable. There, as in other tropical regions, birds are almost as important for pollination as insects, and some scientists go so far as to say that, in fact, they have been the creators of the colorful host of tropical flowers in the long, long process of evolution. Some South American plants depend on hummingbirds for their pollination to such an extent that they fail to produce good seeds when grown elsewhere. I have a feeling that this must be the case with the beautiful red-flowered hibiscus, *Hibiscus rosa-sinensis,* which is cultivated widely in Indonesia, but which I have never seen in fruit there. The Mexican century-plant, whose pollen is carried by hummingbirds, also remains barren when transplanted to Europe—this in spite of the fact that the flowers are abundantly visited by bees there.

Now a few figures, just to give some idea of the importance of birds for our story. There are about 2,000 species of birds, belonging to about 50 families, that visit flowers more or less regularly. About two-thirds of these birds are specialists, relying on flowers as their most important, or even their sole, source of food. In America, such specialists are the hummingbirds (Trochilidae), on the Hawaiian Islands the honey-creepers (Drepanidae). In the region of Australia and New Guinea we find such queer customers as the honey-eaters (Meliphagidae) and the brush-tongued parrots (Trichoglossidae); in Africa and Asia, finally, there are several groups of birds with flower-loving representatives such as the nectar-birds (Nectariniidae) and the spectacle-birds (Zosteropidae).

Of course, we can also approach the problem from the plant side. All in all, there are about 300 families of flowering plants, and a good one-third of these have at least *some* members with flowers that appeal to birds (the so-called "ornithophilous flowers").

As to the history of the case, the first scientist to record visits of birds to flowers probably was Georg Everhard Rumphius, that stouthearted blind seer and Moluccan pioneer who has left us such a treasure-trove of information in his *Amboins Kruidboek.* He mentions that in the Moluccas, the spice islands in the eastern part of Indonesia, certain parrots come to the coral-tree, *Erythrina indica,* to drink the "dew" (the nectar) which accumulates in the flowers. In 1731 Mark Catesby, in his *Natural History of Carolina, Florida and the Bahama Islands* recorded regular visits of certain

American birds to flowers. However, neither he nor Rumphius understood the connection with pollination (of course not, since this
was before Sprengel's day). It was not until 1874 that these things
were seen in the proper light. In that year, Th. Belt described a

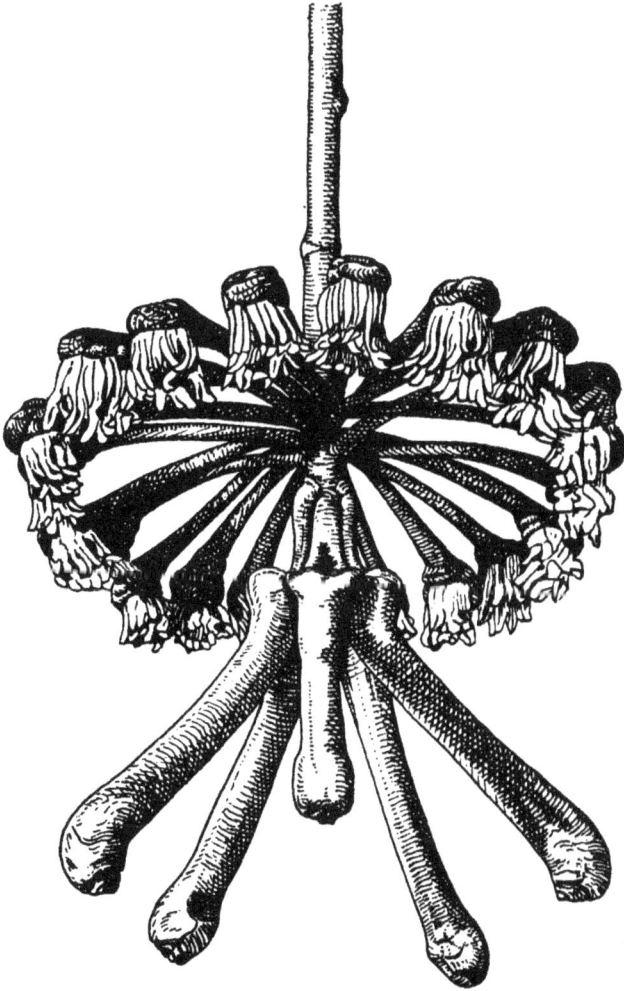

Figure 37. Inflorescence of a *Marcgravia* species from Central America with
five large nectar containers at the base. In all likelihood, birds are the normal
pollinators in *Marcgravia*, although bat-pollination is suspected in a few cases.

genuine hummingbird-flower, *Marcgravia nepenthoides*, in his well-
known book *The Naturalist in Nicaragua* (compare Fig. 37). This
triggered a steady flow of observations and minor discoveries. It is
nice to know that one of these was made by a 5-year-old, Hans
Lorenz, who was the famous Fritz Müller's grandson. In 1886,

this youngest biologist of all time discovered that the flowers of the Brazilian feijoa tree, a *Eucalyptus* relative, are regularly visited by small *Thamnophilus* birds, because these like to eat the four sweet, fleshy petals which in this case are rolled up to form a tasty omelette.

Birds have powerful vision but a very poor sense of smell. It stands to reason that a flower which is attractive to them must have characteristics which are different from those of a bee-flower, for example. Yet it is hard to put our finger on one special, single type which we could describe as *the* bird-flower. The reason is that birds are highly developed animals which behave in a much more "intelligent," or at least in a less rigid, way than do insects. We must also keep in mind that flower-loving birds vary considerably in size.

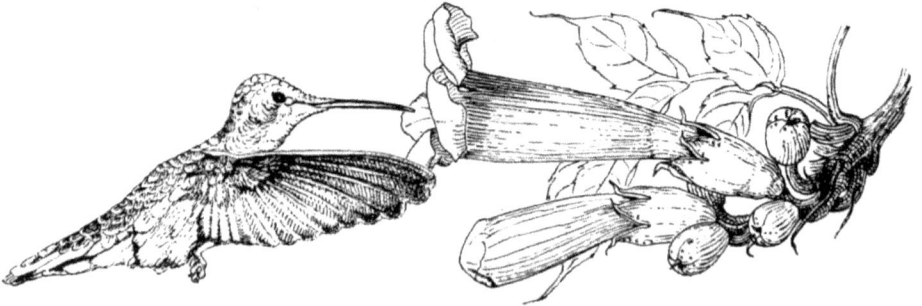

Figure 38. Hummingbird hovering in front of a trumpet-vine flower.

To be sure, most of them are quite small and some do not weigh more than a dime, but others are almost as large as a small crow. Bird-flowers vary accordingly, although most of them are good-sized. Their colors are lively. Quite often, they are orange, red, or scarlet; we may mention cardinal flower, red currant, fuchsia, red columbine, and trumpet-vine (Fig. 38). However, fantastic color combinations which give the flowers the appearance of tropical parrots are not rare, either. Many, or perhaps most, bird-flowers are odorless. In this connection it is very illuminating to compare the sweet-smelling and light-colored European honeysuckles, which are pollinated by hawkmoths, with the completely odorless and orange-red honeysuckle, *Lonicera ciliosa,* of our Pacific Northwest, which is bird-pollinated.

There is an abundance of thin nectar in bird-flowers, and it is protected from dropping out in very ingenious ways, for example, by neatly arranged hairs and ridges. Some scientists claim that bird-pollination got a start in hot climates when the birds began to visit flowers to quench their thirst. Certainly the quantity of nectar produced by Rumphius' coral-trees is very impressive. Each single flower contains almost a thimbleful, and in the dry season each tree is, for months on end, covered with countless blooms. A certain South African plant that caters to birds is known as "suikerbossie" or sugarbush.

Bird-flowers often show protections against damage that might be done by a bird probing into the floral chamber with its sharp-pointed bill. For instance, the nectar is sometimes put in a special spur, separate from the main flower tube, or the ovules are put out of harm's way in an ovary under the floral chamber.

The stamens of coral-tree flowers look like pieces of thick, yellow wire and are almost as strong and tough. In general, we can say that the stamens of bird-flowers are brightly colored and numerous; they stick out so that they touch the bird on the head or breast while it feeds. A few bird-flowers have explosion mechanisms which powder the visitors liberally; others, such as the fancy and colorful bird-of-paradise flower (*Strelitzia*), have pollen grains that stick together because they are connected by threads. Thus, one single pollination visit may be enough to ensure the fertilization of hundreds of ovules. The position of the pistil is just right for this, too.

Unbidden guests are kept out of bird-flowers in a variety of ways, but most often simply by refusing them the welcome mat. This is very striking in an African plant, a species of *Leonurus*, which belongs to the mint family. Whereas most of its relatives have flowers with a very conspicuous lower lip serving as a landing platform for bees (hence their name, Labiatae), there is no such thing at all in the bright-orange flowers of this *Leonurus*.

Hummingbirds usually suck the nectar of flowers on the wing, and in connection with this their favored flowers are often of the hanging type—fuchsias, for instance. In Asia, the situation is somewhat different, because the flower-loving birds there are not such good flyers and have to be provided with some sort of perch in the form of branches or twigs. As we have already seen in an earlier chapter, difficulties arise when certain American fuchsias begin to spread and flourish in the mountains of Indonesia. The oriental birds are at a loss about what to do and end up burglarizing the flowers in the manner of the bumblebees mentioned earlier. However, since the chance of my readers' seeing Asiatic flower birds in action is only slight, I should really concentrate on hummingbirds. They are found exclusively in America, from Tierra del Fuego in the south of Patagonia to Alaska in the north, and at elevations up to 15,000 feet in the Andes. Their bill is thin, very sharp, and usually a little curved. It is really a marvelous tool for sucking up nectar. The upper and lower halves form an airtight tube, in which the forked tongue can be stuck out just about as far as in the woodpeckers, who are famous for this.

Hummingbirds are phenomenal flyers. Their wings whir up and down at such an incredible rate (3,300 times a minute!) that they are just a blur when the animal is hovering. Hummers pollinate flowers in much the same way hawkmoths do and have often been confused with them by hunters. They are so active as flower visitors and sippers of nectar because they are warm-blooded animals with a high metabolism. Each hummingbird has to visit thousands of flowers a day to get enough fuel for the wonderful little engine of its body.

We now have some very good information on a hummingbird's performance, thanks to a beautiful piece of work by Oliver Pearson who points out the following: The smaller an animal, the faster its metabolism, that is, the faster it burns up its body substances with the aid of oxygen. If the cells of an elephant had the same intense metabolism as those of a mouse or a shrew, the bulky animal would not be able to get rid of the resulting heat rapidly enough, and would give up the ghost within a few minutes from overheating. The minute hummingbirds, as we might expect, have the highest rate of metabolism of any bird or mammal. In a resting hummingbird, each gram of tissue metabolizes more than 100 times as fast as a gram of average elephant tissue. A hummingbird, then, is like a Roman candle which burns itself up at a tremendous rate, and if the animal did not have some special tricks up its sleeve which are unknown to elephants, it simply could not eat fast enough to avoid death from starvation. As things are, hummers must consume enormous quantities of food, anyway. A good deal of their day is devoted to the gathering of nectar and insects. During the night, however, these "visual feeders" naturally cannot continue to eat, and if their intense metabolism should go on unabated as it does in other birds, they would not live to see another day. The hummingbird's trick to avoid this dire fate is *to go into hiber-*

*nation every evening.* His metabolic rate then drops sharply, so that by the middle of the night it is only one-fifteenth of the daytime rate. The hummer's temperature now is not much higher than that of the surrounding air (say, 75° F), and the animal is so torpid that it can be picked off its perch like a ripe fruit. Before daybreak, the bird's body temperature spontaneously begins to climb again, the rate of metabolism increases, and when the new day comes, the hummer is its old self again, warm and alert.

This trick of hibernation, by means of which hummers stretch their food stores from dusk to dawn, is unique among the birds. Among mammals, bats are an example of it, but they carry it out in reverse, hibernating by day and feeding by night. Shrews, which are very close to hummingbirds in size and have a rate of metabolism which is only slightly lower, do not know the trick. For this they must pay the penalty of remaining busy most of the night, feeding themselves.

A hovering hummingbird uses up about six times as much oxygen as a resting one. It is interesting to find that its energy consumption compares favorably with that of a modern helicopter, being close to 750 British thermal units per pound per hour. Flying in a forward direction at the high cruising speed of migrating hummers, 35 miles per hour, should not require much less energy. Knowing this, one can calculate how much reserve fat a ruby-throated hummingbird would have to use up in order to make a non-stop flight from its winter area in Central America, across the Gulf of Mexico, and back to the United States. The outcome, according to Pearson, is disappointing; there just is not enough fat in a hummer to do it! The shortest distance across the Gulf is more than 500 miles, and by burning up all its fat as fuel a rubythroat could cover only 385. For this reason, we do not yet know how to explain reports that flights of hummers in the spring head out over the Gulf. On the other hand, the idea that they follow the coast line does not appeal to many scientists, either. Among other things, this would involve a great loss of time and effort for the birds. Whatever the explanation may turn out to be, let us gratefully accept the fact that the rubythroats *do* come back to the eastern United States, and the rufous hummingbirds to our Pacific Northwest. Could anyone ask for a more delightful pair of spring harbingers than rufous hummingbird and red-flowering currant?

# 11

# Busy Bees and Efficient Bumblers

Somebody has estimated that there are at least 5,000 species of bees in North America; on the flowers of alfalfa alone, more than 100 species have been recorded as visitors. However, it is not necessary to know them all to get a lot of fun out of the bees. Let us try to learn just a few of the most important ones. We shall start slowly, early in the spring, a time when there are not too many bees around. This is really wise strategy, for (as we will soon find) the different types do not emerge at the same time so that we can tackle them one by one.

The easiest bees to get to know are, of course, the honeybees (Fig. 39). They are pampered pets that spend the winter in a hive but will come out very early in the year, provided the weather is sunny enough. It is a good thing that so many people nowadays grow Christmas roses and Lenten roses (*Helleborus* species) in their garden, for these produce a lot of nectar and are really a great help to our early "birds." Even a superficial inspection shows that *Helleborus* flowers have nothing to do with genuine roses; they are closely related to the buttercups and anemones. The nectar is offered to the bees in delightful, miniature pitchers, to be found between the "petals" and the bunch of stamens that occupies the center of the flowers (Fig. 28, page 78). The only thing the honeybees have to do to gorge themselves is to stick their heads into these little cornucopias.

There is very little competition for honeybees from other bees, so early in the season, but I must warn you that a certain fly which is a dead ringer for our honeybee is already around. This is the dronefly, *Eristalis taenax*. However, with a little bit of attention one can tell the two insects apart easily. It is a general rule that a fly has only two wings, a bee four. Our customers are no exception. You may also notice that, like many other flies, *Eristalis* has the habit of "rubbing its hands together" occasionally. I hope that

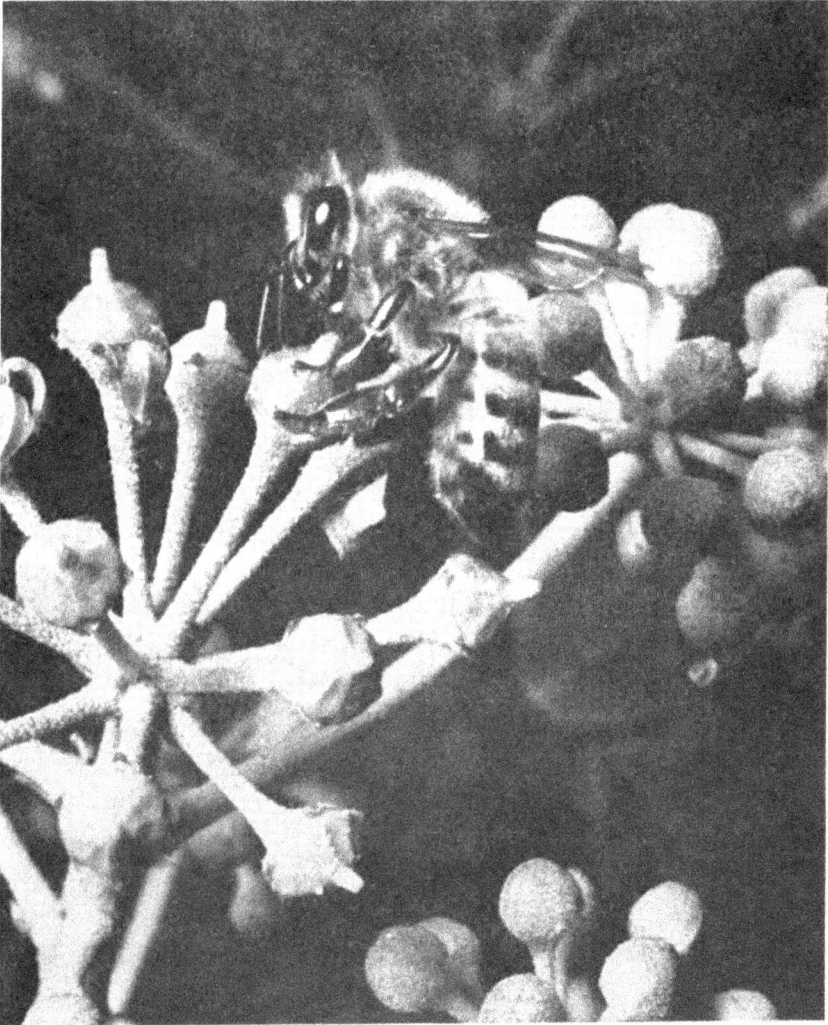

Figure 39. Honeybee collecting nectar on the flowers of English ivy, *Hedera helix*.

you will watch both insects so closely that they will leave a lasting imprint on your mind, for they are fascinating, so much so that I will devote a whole chapter ("The Scarab's Sister") especially to the honeybee and another one ("The Imitators") to *Eristalis* and its close relatives.

Apart from the honeybee, practically all bees and bumblebees hibernate in a state of torpor. Occasionally, you may come across one or two bumblebees in the cold season, when you are turning over sods in your garden, but you have to be a really keen observer to see them at all. They keep their wings and feet pressed tightly against their bodies, and in spite of their often colorful attire you may very well mistake them for lumps of dirt. I must add at once that these animals are what we call "queens," young females that have mated in the previous summer or autumn. It is on them alone that the future of their race depends, for all their relatives (mothers, husbands, brothers, and unmated sisters) have perished with the arrival of the cold weather. Even some of the queens will die before the winter is over, falling prey to enemies or disease. The survivors emerge on some nice, sunny day in March or April, when the temperature is close to 50° F and there is not too much wind. Now the thing for us to do is to find ourselves a couple of those wonderful flowering currants such as the red *Ribes sanguineum* of our Pacific Northwest, or otherwise a good sloe tree, or perhaps some nice pussy willow in bloom, preferably one with male or staminate catkins. The blooms of *Ribes* and of the willow and sloe are the places where large numbers of our early insects will assemble: honeybees, bumblebees, and other wild bees, and also various kinds of flies. It is a happy, buzzing crowd.

Each male willow catkin is composed of a large number of small flowers. It is not difficult to see that the stamens of the catkin are always arranged in pairs, and that each individual flower is nothing but one such pair standing on a green, black-tipped little scale. By scrutinizing the flowers, one can also notice that the scale bears one or two tiny warts. Those are the nectaries or honey glands (Fig. 26, page 74). The staminate willow catkins, then, provide their visitors with both nectar and pollen; a marvelous arrangement, for it provides exactly what the bee queens need to make their beebread, a combination of honey and pollen with which the young of all species are fed. The only exception to this is certain bees that have become parasites. I will deal with these later on.

Quite often, honeybees form a majority on the willow catkins. As we have already seen in the first chapter, bumblebees are bigger, hairier, and much more colorful than honeybees, exhibiting various combinations of black, yellow, white, and orange. Let us not try to key them out at this stage of the game, and let us just call them *Bombus;* there must be several dozen species in the United States alone. If you really insist on knowing their names, an excellent book on the North American species is *Bumblebees and Their Ways* by O. E. Plath.

If we manage to keep track of a *Bombus* queen after she has left her feeding place, we may discover the snug little hideout which she has fixed up for herself when she woke up from her winter sleep. As befits a queen, a bumblebee female is rather choosy and may spend considerable time searching for a suitable nesting place. Most species seem to prefer a ready-made hollow such as a deserted mouse nest, a bird house, or the hole made by a woodpecker; some show a definite liking for making their nest in moss. Once she has made up her mind, the queen starts out by constructing, in her chosen abode, a small "floor" of dried grass or some woolly material. On this, she builds an "egg compartment" or "egg cell" which is filled with that famous pollen-and-nectar mixture called beebread. She also builds one or two waxen cups which she fills with honey. Then, a group of eggs is deposited in a cavity in the beebread loaf and the egg compartment is closed. The queen afterward keeps incubating and guarding her eggs like a mother hen, taking a sip from time to time from the rather liquid honey in her honey pots. When the larvae hatch, they feed on the beebread, although they also receive extra honey meals from their mother. She continues to add to the pollen supply as needed.

The larvae, kept warm by the queen, are full grown in about ten days. Each now makes a tough, papery cocoon and pupates. After another two weeks, the first young emerge, four to eight small daughters that begin to play the role of worker bees, collecting pollen and nectar in the field and caring for the new young generation while the queen retires to a life of egg laying. The first worker bees do not mate or lay eggs; males and mating females do not emerge until later in the season. The broods of workers that appear later tend to be bigger than the first ones, probably because they are better fed. By the middle of the summer, many of the larvae apparently receive such a good diet that it is "optimal," and it is then that young queens begin to appear. Simultaneously, males or drones are produced, mostly from the unfertilized eggs of workers, although a few may be produced by the queen. The young queens and drones leave the nest and mate, and after a short period of freedom, the fertilized young queens will begin to dig in for the winter. It is an amazing fact that in some species this will happen while the summer is still in full swing, for instance, in August. The temperature then is still very high. At the old nest, the queen will in the early fall cease to lay the fertilized eggs that will produce females. As a result, the proportion of males (which leave the nest) increases, and eventually the old colony will die out completely. The nest itself, the structure that in some cases housed about 2,000 individuals when the season was at its peak, is now rapidly destroyed by the scavenging larvae of certain beetles and moths.

Not always, though, does the development of a bumblebee colony take place in the smooth fashion we have just described. Some members of the bee family have become idlers, social parasites that live at the expense of their hardworking relatives. Bumblebees can thus suffer severely from the onslaughts of *Psithyrus,* the "cuckoo-bumblebee" as it is called in some European countries. Female individuals of *Psithyrus* look deceptively like the workers and queens of the bumblebees they victimize. The one sure way to tell victim and villain apart is to examine the hind legs which in the case of the idler, *Psithyrus,* lack the pollen baskets—naturally! The female parasite spends much time in her efforts to find a nest of her host. When she succeeds, she usually manages to slip in unobtrusively, to deposit an egg on a completed loaf of beebread before the bumblebees seal the egg compartment. The hosts never seem to recognize that something is amiss, so that the compartment afterward is sealed normally. Thus, the larvae of the intruder can develop at the expense of the rightful inhabitants and the store of beebread. Later on, they and the mother *Psithyrus* are fed by the *Bombus* workers. Worse still, in a number of cases it has been claimed that the *Psithyrus* female kills off the *Bombus* queen.

But let us return, after this gruesome interlude, to our willow catkins in the spring; there are other wild bees that command our attention.

It is almost certain that some of these, usually a trifle smaller than the honeybees, are andrenas or mining bees. There are about 200 different kinds of *Andrena* in Europe alone. One of my favorites is *A. armata,* a species very common in England, where it is sometimes referred to as the lawn bee. The females like to burrow in the short turf of well-kept lawns, where their little mounds of earth often appear by the hundreds. Almost equal in size to a honeybee, *A. armata* is much more beautiful in color, at least in the female of the species: a rich, velvety, rusty red. The males are much duller.

After having mated, an *Andrena* female digs a hole straight down into the ground, forming a burrow about the size of a lead pencil. The bottom part of a burrow has a number of side tunnels or "cells," each of which is provided with an egg plus a store of bee-bread. The development of the *Andrena* larvae is very rapid, so that by the end of spring they have already pupated and become adults. But they are still enclosed in their larval cells and remain there throughout the summer, fall, and winter. Their appearance, next spring, coincides in an almost uncanny way with the flowering of their host plants. In the Sacramento valley in California, for instance, it has been observed that there was not one day's difference between the emergence of the andrenas and the opening of the willow catkins. This must be due to a completely identical response to the weather, in the plant and the animal.

After the male and female andrenas have mated, the cycle is repeated. Although *Andrena* is gregarious, so that we may find hundreds and hundreds of burrows together, we must still call it a solitary bee. Its life history is much simpler than that of the truly colonial bumblebees and can serve as an example of the life cycle of many other species. After all, social life in the group of the bees is by no means general, although it certainly is a striking feature. On the basis of its life history, we like to think that *Andrena* is more primitive than the bumblebees. The way in which it transports its pollen is not so perfect, either. It lacks pollen baskets and possesses only a large number of long, branched hairs on its legs, on which the pollen grains will collect. Still, *Andrena* will do a reasonably good job, so that an animal with a full pollen load looks like a gay little piece of yellow down floating in the wind.

Closely related to the andrenas are the nomias or alkali bees. *Nomia melanderi* can be found in tremendous numbers in certain parts of the United States west of the Great Plains, for example, in Utah and central Washington. In the United States Department of Agriculture's *Yearbook of Agriculture, 1952*, which is devoted entirely to insects, George E. Bohart mentions a site in Utah which was estimated to contain 200,000 nesting females. Often the burrows are only an inch or two apart, and the bee cities cover several acres. The life history of the alkali bee is similar to that of *Andrena,* but the first activity of the adults does not take place until summer, and the individuals hibernate in the prepupal stage. In most places, there are two generations a year, a second brood of adults appearing late in the summer.

I must plead guilty to a special sympathy for nomias. This may just be pride in my adopted State of Washington, but certainly I love to visit their mound cities near Yakima and Prosser in July or August, when the bees are in their most active period. The name "alkali bee" indicates that one has to look for them in rather inhospitable places. Sometimes, although by no means always, these are indeed alkaline. The thing is that these bees love a fine-grained soil that is moist; yet the water in the ground should not be stagnant either. They dislike dense vegetation. Where does one find such conditions? The best chance, of course, is offered by gently sloping terrain where the water remains close to the surface and where the air is dry, so that a high evaporation leaves salty deposits which permit only sparse plant growth. I am happy to report that the irrigation of desert-like areas in central Washington which forms part of the gigantic Columbia Basin project has created just the right conditions in many places.

It is certainly worth while to protect the nomias and their nesting-sites, especially in areas where alfalfa is grown for the seed. They simply are the best alfalfa pollinators in western North America, and in this particular case they could not very well be replaced by honeybees. It is not that honeybees have a dislike for alfalfa flowers; on the contrary, they love them for their nectar. As a rule, however, they are not too much interested in the alfalfa *pollen.* They steal the nectar from the flowers through a side entrance and fail miserably to "trip" the alfalfa flowers, that is, to set in motion the explosion mechanism for the release of pollen grains. I do not want to bore you with a lengthy description of that mechanism, so let us just look at Figure 40 and say that it is very similar to the one we described for Scotch broom in the chapter "Unbidden Guests Beware." It is almost as if honeybees have a strong dislike for getting a pollen cloud into their faces. Only accidentally, therefore, will they trip a few of the alfalfa flowers. One solution would be to overstock the alfalfa fields with honeybees, but it is hard to create enthusiasm among beekeepers for such a project—after all, they are primarily interested in honey. Another solution is to bring in fresh honeybees continually and to withdraw those that have learned to avoid the unpleasant explosions, but again this is impractical. So, the answer is to employ native, wild bees. We have already seen that these always need pollen for their bee-bread; they will visit alfalfa flowers especially for the purpose of getting it and will effectively trip the majority of the flowers. And

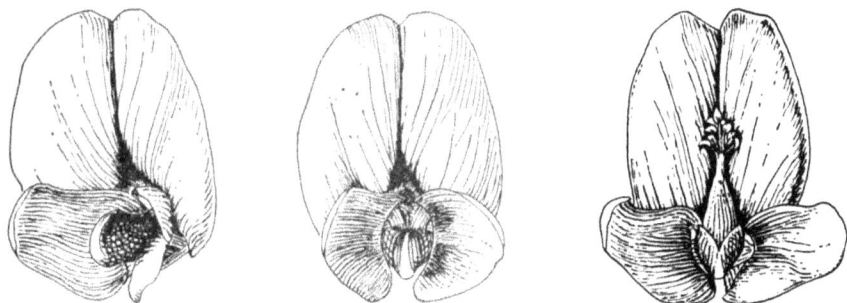

Figure 40. Flower of alfalfa (*Medicago sativa*), before and after "explosion." The column formed by pistil and stamens is originally hidden in, and kept in check by, the keel. It pops up when a bee alights, powdering the animal with a cloud of pollen. Alkali bees (*Nomia*) are more inclined to put up with this than honeybees, which is one reason why they are much better alfalfa pollinators. Honeybees are mostly interested in the honey, which they steal without "tripping" the flowers.

*Nomia* is the only wild bee numerous enough in these areas to do the job. If you want to know more details about these interesting things, you will find them in Bohart's excellent article.

Now that I have started talking about bees that are active in the summer, I should not forget the leaf-cutting bees, *Megachile*, of which there are a great many species. They are mostly dark with rather flattened bodies; some of them are a little larger than honeybees, but most are smaller. The females can be recognized easily enough on flowers by the very funny way in which they collect and transport pollen. Instead of pollen baskets on their legs, they have a brush of hairs, mostly whitish or golden, which covers most of the underside of their bellies. From time to time one can see how a female rubs her belly across the surface of, say, a dandelion flower or daisy, so that her back appears to be quite hollow. This helps in collecting the precious grains. Moreover, the females have

the habit of cutting more or less circular or elliptic pieces out of
leaves, birch bark, or flower petals, always starting from the margin
(Fig. 41). Holding such a piece between her hind legs, a female
will transport it to her subterranean burrow or to the wood cavity
which she inhabits and will skillfully fit it together with other

Figure 41. Leaflets of a rose plant, showing the work of a leafcutter bee,
*Megachile.* Circular and oval pieces are cut out to make the little subterranean
container for the beebread, a mixture of nectar and pollen intended to feed
the *Megachile* larvae. In the case of the smallest leaflet shown here, the leafcutter was disturbed before it could finish its job.

pieces to form a number of thimble-shaped cells (Fig. 42). These
are filled with the usual fare for the larvae, beebread, and are provided with one egg each. After this, the cells are closed with a lid
which again consists of circular pieces of leaves. You will notice
the presence of these leafcutters in your garden easily enough; after
all, when they are transporting their "wallpaper" they are very
conspicuous, especially when the material happens to be a fiery red
geranium leaf! If you do not catch them in this act, you may at
least find the traces of their work in your flowerbeds, on your lilacs,
or on the wisteria growing on your porch. Our figure shows rose
leaves that have served one of our little paper hangers several times.

In some regions of Canada, leafcutter bees are credited with a
major share in the pollination of alfalfa.

In order to be at least somewhat complete, I have to introduce
still another group of bees: the anthophoras or flower-loving bees.
In Holland, we gave them the common name of sachem bees because

of the bundles of long hairs they have on their legs, resembling the feather garb of Indian chiefs, sachems. Anthophoras are long-tongued bees, able to get the nectar out of "difficult" flowers that baffle even some of the bumblebees. In general, therefore, the spring anthophoras are very active on the same species of flowers that are frequented by the long-tongued bumblebees, such as *Daphne*, prim-roses, and *Corydalis*. But when June comes around, it is hard to find

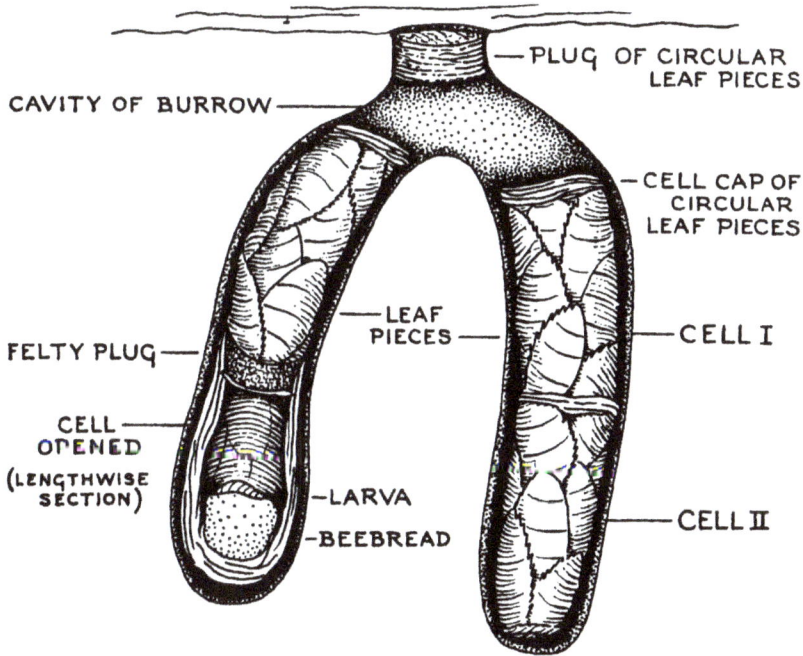

Figure 42. Nest of a leafcutter bee, Megachile.

a single specimen left. In the United States we are fortunate enough to have some anthophoras that are active in midsummer, for example, *Anthophora occidentalis* whose burrows we see in Figure 43. On the island of San Juan, close to the famous American Camp of Pig War days, I know a beautiful city or at least a town of *A. occidentalis*, overlooking a driftwood-filled bay. There are hundreds, maybe thousands, of nests in the hard, clayey soil, each topped by a little curved mud chimney. In July, the whole place is buzzing with activity. The favorite food plant of the bees here seems to be the yellow monkey-flower, *Mimulus guttatus*, which we have al-ready described in Chapter 8 in connection with the movement of the stigma lobes.

Before winding up this discussion of bees, I must draw your attention to the giants of the bee world, the true carpenter bees or xylocopas. It is too bad that these are tropical bees, at home especially in Central Africa, South America, and Indonesia. However, in Europe their range extends as far north as Paris, and they

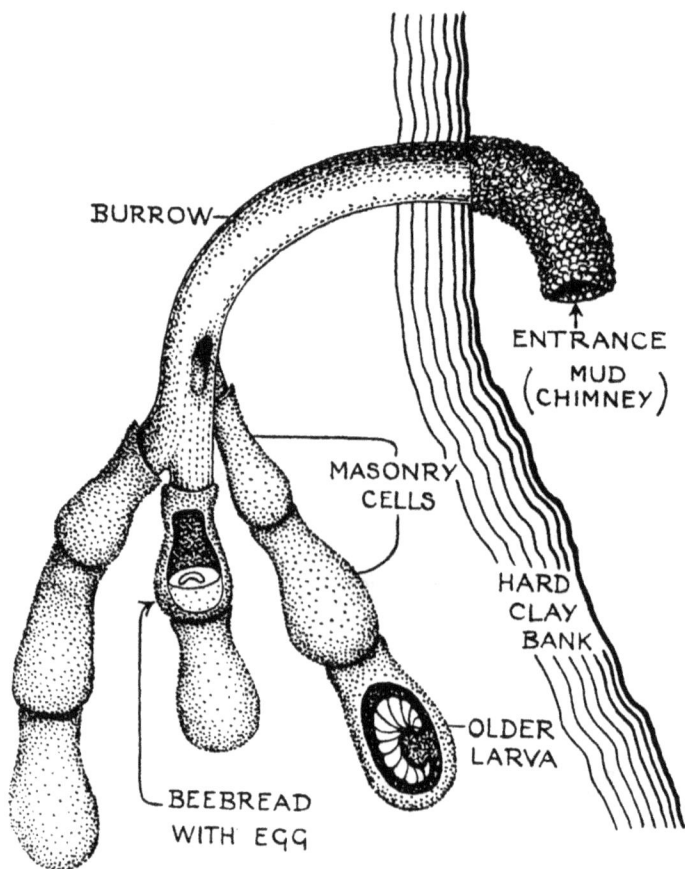

Figure 43. Burrows of *Anthophora occidentalis*. Two cells have been cut open.

are quite common in the bewitching gardens of Cordova and Granada in Spain. At least one species, *Xylocopa virginica*, is found in the southern and southeastern United States. You will recognize it easily, for virginicas are very stout, bigger even than the largest bumblebees, and very dark—at least if you are dealing with females. In this country, P. Rau has made a study of their very interesting habits. I am quite tempted to tell you more about these xylocopas, but it might take us too far afield. Suffice it to say that in many places where bumblebees are lacking, xylocopas will take over their role—as will be shown in the chapter "Ambushes and Traps."

Now a few words about bee-flowers. We have already indicated that, as a rule, they are bright-colored, for instance yellow or blue. They are open in the daytime, have fragrance, and possess a good supply of nectar and/or pollen. Many of them offer the bees a landing platform. In their structure, they may be rather complicated, so that they cannot easily be "worked" by insects other than bees and bumblebees. As a result of all this, there may be a close tie between certain plants and the bees that pollinate them. This is illustrated by Figure 44, showing the distribution on our planet of monkshoods (*Aconitum*) and of bumblebees (*Bombus*). The two areas are almost the same, except for South America.

The only bee-flower we will discuss here at a certain length is *Salvia pratensis*, one of the true sages. It is a plant with dark purplish-blue flowers that is common in western Europe. In a most beautiful way, it displays a combination of independent timing on the part of the flower with a triggering activity on the part of the visitor. A similar mechanism, by the way, is found in the yellow-flowered species *S. glutinosa* and in *S. officinalis*, a plant with pale purplish blooms. All three species can be found in American herb gardens.

The salvias belong to the mint family, the Labiatae. As the family name indicates, the flowers in this group of plants usually have a very distinct upper and lower lip, the latter serving as a landing platform for the bees and bumblebees that are attracted by the bright colors and the fragrance. The animals are very partial to the nectar and pollen which the Labiatae provide in abundance. In the particular case of *S. pratensis*, the upper lip is flattened on the sides, so that it forms a narrow vertical box in which the stamens lie hidden. Of the pistil, which is likewise hidden here, only the upper part with the two stigma lobes is visible.

For all practical purposes, a flower of *S. pratensis* which has just opened is a male flower. The pollen is ripe and ready to be shed, but the stigma lobes are still unreceptive, folded together and sticking out horizontally so that the flower's entrance is left free. A bee or bumblebee alighting on the lower lip and sticking its head in to reach the nectar will activate a peculiar pivot mechanism that will

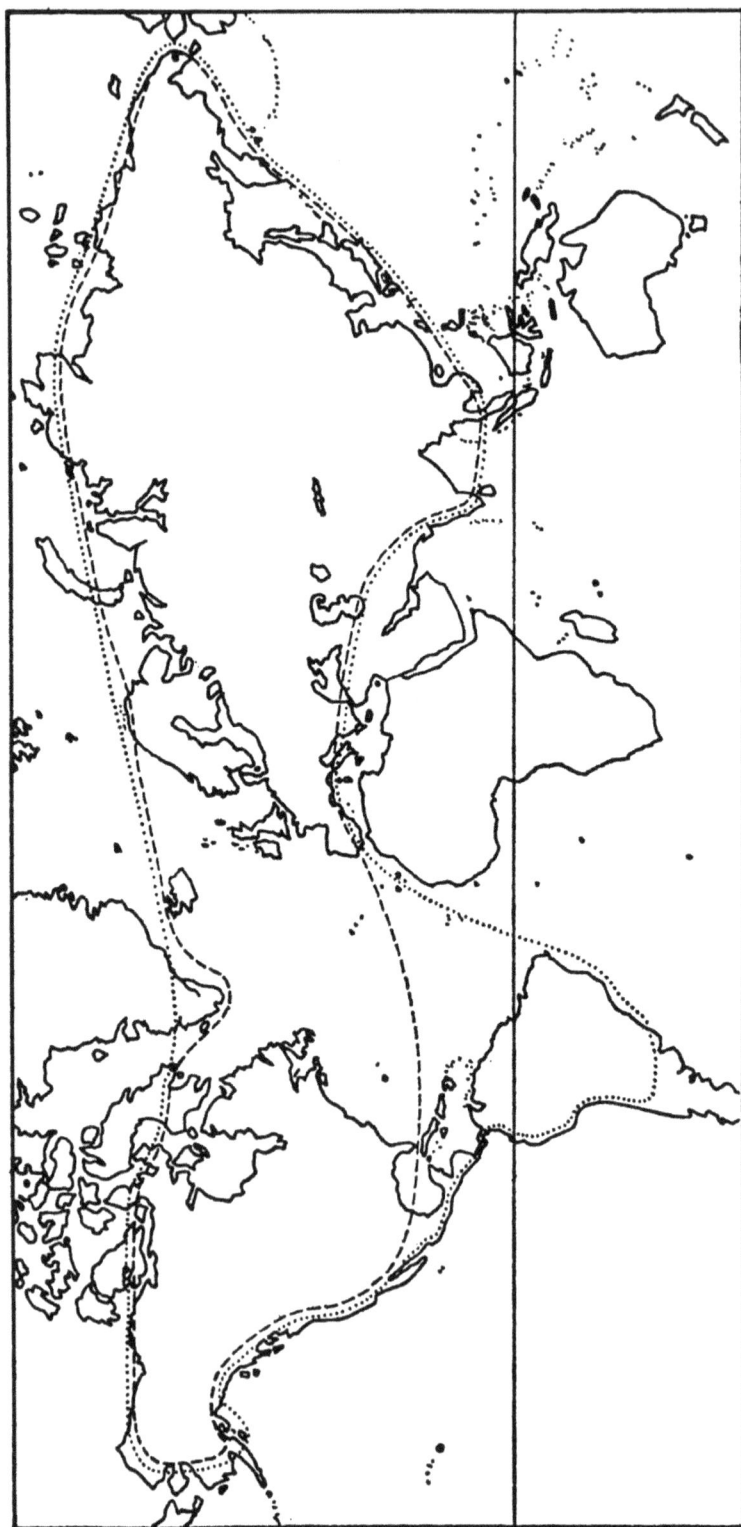

Figure 44. Distribution of the various monkshoods (*Aconitum*) and the bumblebees (*Bombus*). Monkshoods are indicated by the dotted line and bumblebees by the broken line. (After Kronfeld.)

bring the pollen sacs out of their hiding place and down to powder the visitor's back, as shown in Figure 45 which will help us understand this situation. In contrast to most other members of its family, a *Salvia* flower has only two, not four, well-developed stamens. However, even these two are peculiar. Each filament is so short as to be unrecognizable, forming a joint around which the strangely shaped anther can pivot. The lower half of the anther has been transformed into a spoon-shaped, vertical plate and the upper into a drawn-out, long piece that looks very much like a filament. It is the free end of this piece which bears the two loosely attached pollen sacs. A small, rod-shaped object, such as the proboscis of a bee or a piece of wire which penetrates into the flower in a horizontal fashion, will find the basal plate in its way and cannot avoid pushing it back, causing the upper arm of the anther to come out of hiding, as indicated. As soon as the pressure is released, the upper arm with its pollen sacs snaps back into the upper lip, but the visiting bee or bumblebee has already received a dose of pollen on its back.

During the next few days the flower will transform itself into one which, for all practical purposes, is female. The pistil stretches out and curves downward, and the stigma lobes—now receptive—spread out, blocking the entrance to the flower. Visitors who are keen on the nectar (which continues to be secreted) simply *must* touch them, and if the visitors happen to come from a "male" *Salvia* flower which has just dusted them with pollen, Nature's purpose is achieved.

It is very illuminating to compare *S. pratensis* with *S. splendens*, from Brazil, a species that often remains sterile in the temperate gardens where it was introduced, but which is successfully pollinated by hummingbirds in its native country. The color of *S. splendens* is a flaming red, there is no pronounced smell, and the lower lip is not too well developed.

Another comparison is in order, too. The flowers of Jackbean (*Canavalia*), a plant which belongs to the pea family, show the general structure which we can expect in that group, with keel, wings, and banner. As a matter of fact, there is not too much difference from the flowers of Scotch broom illustrated in Figure 19, page 54. However, in contrast to most of their relatives, the Jackbean flowers in Nature have a position with the banner *down*, not

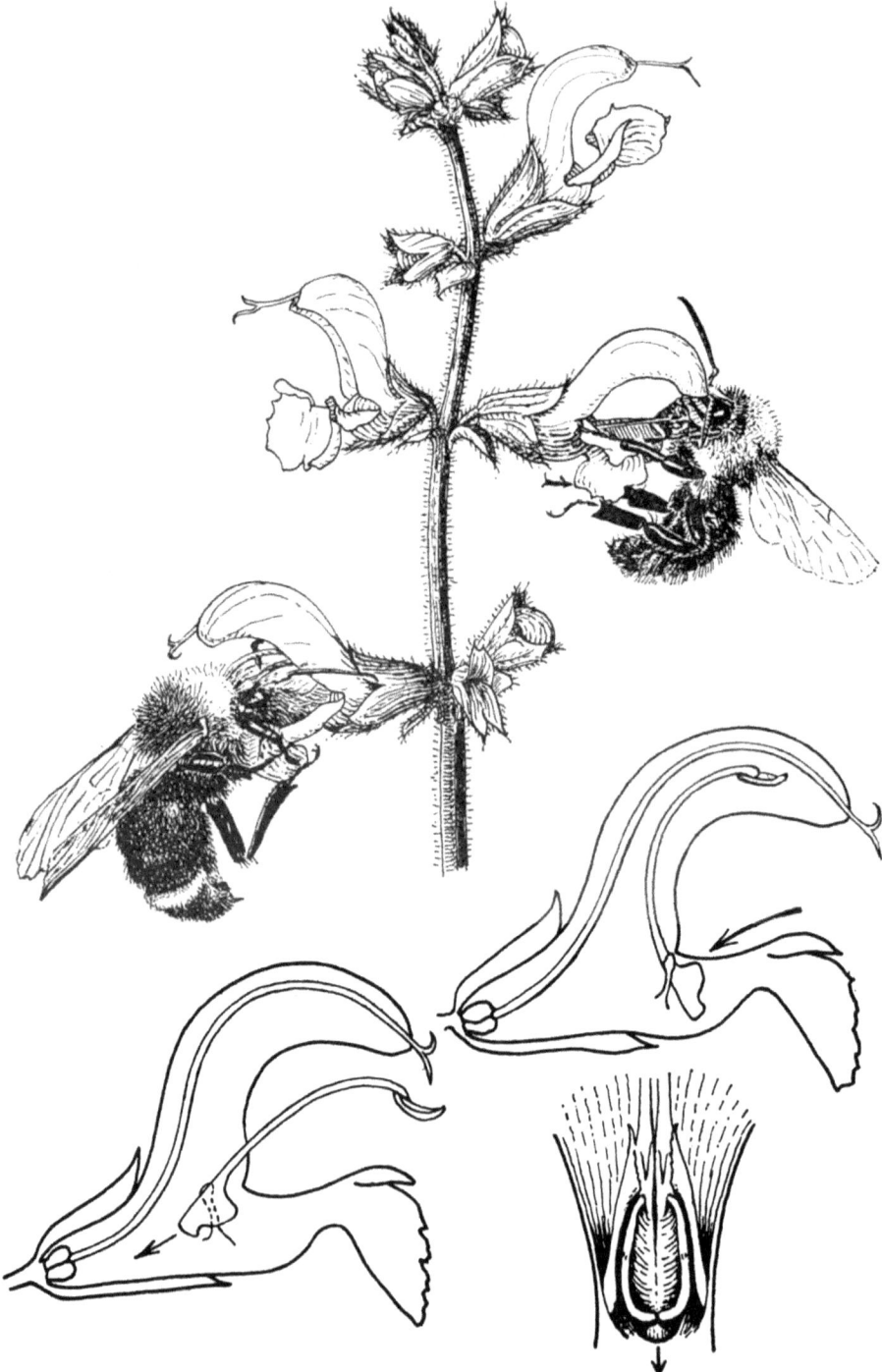

Figure 45. Bumblebees in action on the flowers of the blue meadow-sage (*Salvia pratensis*).

up! They are simply turned upside down. This makes them very excellent imitations of the flowers of Labiatae, and they are "manipulated" by bumblebees and the big tropical carpenter bees (*Xylocopa*) in much the same fashion as the latter. Again, it is the back of the visitor which receives the pollen. One can do very nice experiments with these Jackbean flowers by keeping the flowering spray upside down. This baffles the poor visitors completely and they no longer know what to do.

# 12

# *The Scarab's Sister*

The ancient Egyptians were not only keen observers of animals but also possessed the ability to see divine principles in vulgar, everyday things. The scarab beetle rolling its ball of dung reminded them of the movement of heavenly bodies and was raised to the rank of a sacred animal; the image of the creature was used to replace the heart of the mummified dead (Fig. 46). Likewise, the honeybee was regarded as a symbol for the kings of Lower Egypt. Is it because of its formidable sting that it was so often chosen to represent a monarch powerful in war on the stone slabs of temples and tombs? (Figs. 47, 48, and 49.) Or was it simply because the Egyptians were impressed with the magical ways of the bees? We can only guess, but one thing can be said: had the Egyptians had just an inkling of what we know about bees today, the scarab might well have lost its first place. "Magical" is, indeed, the only way to describe the behavior of hive bees, and it is altogether proper that we devote an entire chapter to them.

Of all the bees, hive bees or honeybees have the most beautifully developed colonies or "states." Although, as we shall discuss later, we sometimes find these in hollow logs or caves, the general rule is that the home of a bee colony is in a hive. Under normal circumstances, there is only one queen there and a few thousand worker bees. The males, or drones, are found only during a period of a few weeks each year. The queen represents the stable element in the state. She lives to the ripe old age of 3 to 5 years, whereas the worker bees die after a few weeks or, at best, months. The bee queen has to pay a price for this longevity, however, and is much more immobile than the bumblebee queens which we mentioned before. All her life, at least during the warm season, she does nothing but lay eggs, usually two every minute, for a lifetime total of as many as a million. Each day, she produces more than her own body weight in eggs! This is possible only because the queen is at all times surrounded by a throng of workers who keep feeding her.

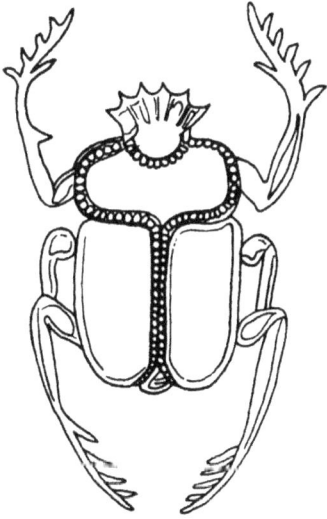

Figure 46. Scarab from a sarcophagus in the Louvre, Paris, twenty-sixth Egyptian dynasty, about 600 B.C. (After Carpenter, by permission.)

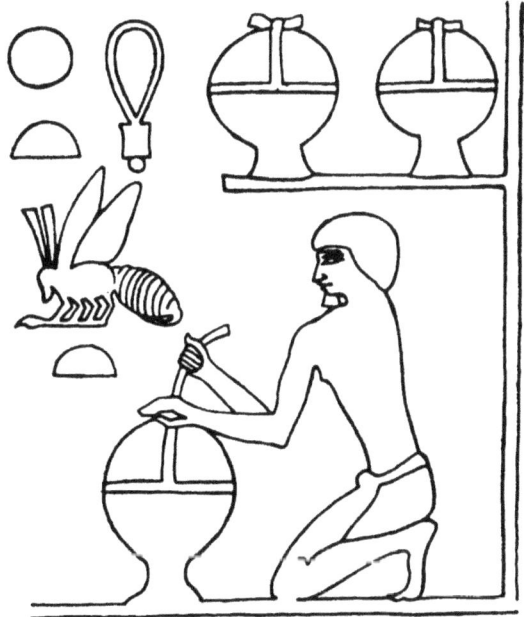

Figure 47. Beekeeping in ancient Egypt. In addition to the bee, we see a slave sealing a honey pot, some utensils, and more honey pots. (After Newberry.)

In the life of these worker bees, which is by no means so monotonous as that of the queen, three main periods can be distinguished. In the first ten days after they have hatched, worker bees act as charwomen and nursemaids. For instance, from the third to the fifth day of her adult life, a worker bee has the job of feeding pollen and nectar to older larvae; from her sixth to her tenth day, she feeds younger ones a nutritious juice from her salivary glands. For this purpose she will, in this period, eat quite a bit of protein-rich pollen, which is brought to the hive by older workers.

When these first ten days are over, a worker bee will go out on her first flights. As might be expected, the very first one is for orientation only. The salivary glands now gradually lose their activity, but the wax glands, which are to be found on the underside of the abdomen and are connected with the outer world by minute pores, reach complete development. Now it is the task of the workers to build combs and to close with a waxen lid the cells which contain honey or larvae. They produce the wax in the form of small plates which emerge from the abdomen. The animals grab hold of these with their legs, put them in their mouths, and chew on the wax for a while. As a result of this operation, the substance becomes fit for building. There are three types of cells

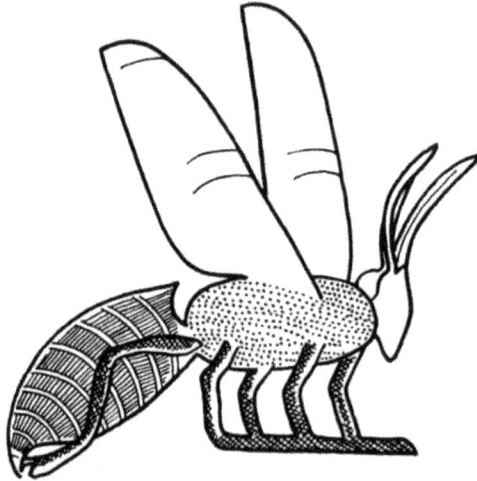

Figure 48. Ancient Egyptian picture of a bee in baked clay, fourth dynasty, beginning of the Old Kingdom, 2800–2400 B.C. (After Carpenter, by permission.)

containing larvae. Those from which, later on, worker bees will emerge are smallest, and the drone cells are a trifle larger. The cells for the future queens are very much bigger, with thicker walls and quite a different shape; they lack the beautiful "mathematical" form of the ordinary cells and are only roughly cylindrical. Also, they are suspended separately from the comb. It is very much worth your while to observe a natural bee comb leisurely and to convince yourself that the particular shape of most cells guarantees a very economical use of that precious material, wax. In comparison, bumblebees act in a way that can only be described as happy-go-lucky. As we have seen, they just throw together a number of very irregular cells, with an almost admirable disregard for economy.

After a short transition period, in which the worker bees act as gatekeepers, they are promoted to the rank of foragers. It is now their task to collect nectar and pollen in the field, guided by colors and nectar-guides in the manner already described. One such food-gathering expedition may last as long as 2 hours, and may cover a distance of from 2 to 3 miles. Upon their return, the bees pass their nectar on to a young worker bee, or they regurgitate it into cells reserved especially for that purpose. They will also rub off, into special cells, the pollen which they brought home in their pollen baskets. In a few minutes, they are ready for a fresh start again.

Figure 49. A cylinder seal of King Amenemhat III, with bee. Egypt, 1849–1801 B.C. (After Newberry.)

If a bee happened to return from a very rich food source, she will not fly out again before having performed a "food dance," which communicates the position of the food source to other foragers waiting in the hive! There are two main types of food dance: a simple "round" dance and a "figure-of-eight" dance which is accompanied by peculiar tail-wagging movements (Fig. 50). The round dance simply indicates that there is an abundant food source with a radius of about 100 meters, that is, a little more than 100 yards, from the hive. But of course the dancing bee will, automatically, give other information as well. After all, the peculiar scent of the flowers she visited will be on her body and also in the nectar or pollen which the waiting foragers in the hive receive from her. Finally, the scent with which the finder marked her treasure, in the good old tradition of pioneering forager bees, will also be detected by the would-be recruits. All in all, then, there is very little chance that the food source will not be found rapidly by a whole crowd of excited foragers. However, for food sources farther away than 100 meters, more information is required, and this is provided by the figure-of-eight dance. Both the direction and the distance of the food source are indicated. This is indeed a magnificent achievement, perhaps even a unique one in the animal kingdom. I do not believe that there are many animals besides man that can thus point and send their colleagues almost infallibly to a faraway spot. The only example that comes to my mind is that of the female eider ducks, who will point out to their husbands other males that should be chased away. However, the distances involved in this case are miserably short.

Von Frisch, who discovered the almost magical communication methods of honeybees, applied the very neat trick of watching them through the glass wall of an observation hive in red light, to which the bees, as we have seen, are not sensitive. Under such an illumination, the animals continue their dances, which would other-

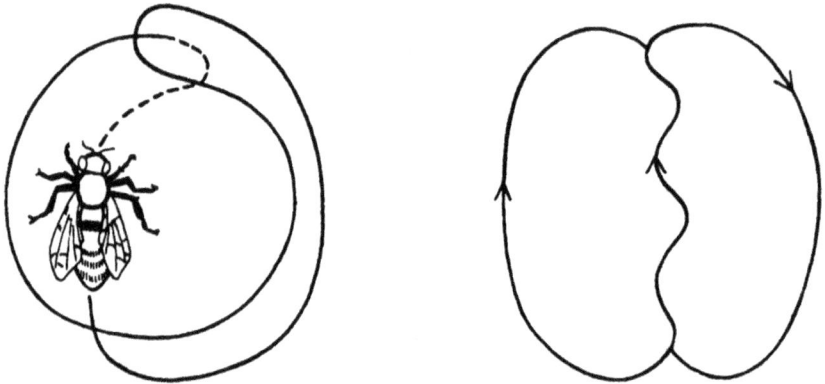

Figure 50. Dance figures of the honeybee. (After von Frisch.)

wise be performed on the vertical face of the comb in utter darkness. The number of tail-wags which the bee makes as she moves along the middle of the figure-of-eight indicates distance. If she waggles her abdomen quickly, it means that the food source is close by; if she does it more slowly, the food is far away. For a food source that is about a mile away, we would normally see about four wags in each period of a quarter of a second. "Distance" here is not the precise distance, or distance "as the crow flies," but something much more useful to the bees. If the dancing bee had to battle against a head wind in her flight toward the food, she will very neatly indicate by her wags a greater distance than the one she really covered. In other words, she adds something for "effort." Likewise, if she was helped by a tail wind, she will indicate a shorter distance. The direction of the food source in relation to the sun is indicated by the direction of the bee's dance on the comb. For instance, if the middle part of her figure-of-eight (the part where she wags her tail) is exactly vertical and she is going *up*, here, she means that the foraging bees must head straight toward the sun. If it is vertical and *down*, they must head away from the sun to reach the food source. A dance at an angle to the vertical of, say, 60 degrees to the left, indicates that the bees must follow a direction which forms the same 60-degree angle with the line connecting the hive and the sun. All this sounds rather complicated, but I hope Figure 51 will make things clear.

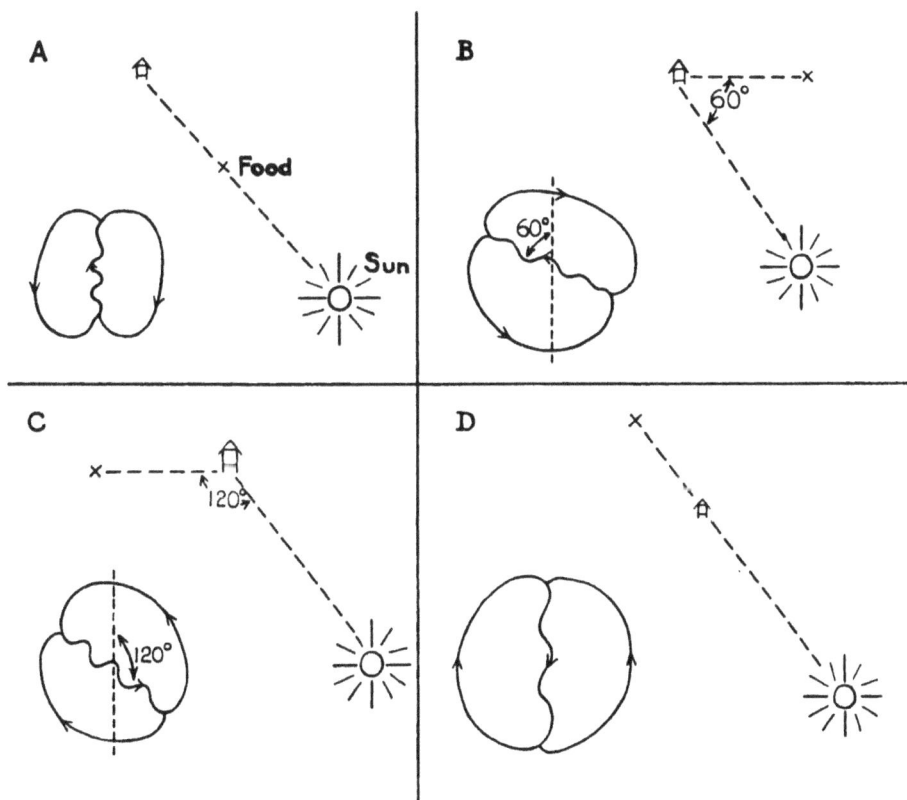

Figure 51. Indication of a food source by the honeybee, with the sun as a beacon. (By permission of Cornell University Press.)

The system works very well, so that it is indeed difficult to fool the bees. But "height" is something they cannot indicate in their dance, and therefore there is one fine way to baffle them completely. We simply put their hive at the bottom of a radio mast and haul the food to the top of it, directly above the hive. Although some foragers will find the food, they cannot "tell" other bees about it in their dance language, and move in crazy patterns, just as if they were raving or stuttering. Of course, this is a most "unfair" experiment because we confront the bees with a situation which they are most unlikely to encounter in their normal lives.

One of the most amazing things is that the bees will still indicate, and find, food sources correctly even when they cannot see the sun directly. How is this possible? The answer is that they pay attention to the *polarization* of the light waves which reach them from different parts of the sky. This requires a few words of explanation.

When we move the end of a horizontal rope up and down quickly, we can produce a number of waves which travel along the rope, each of them vertical in direction. What we call "light" can be thought of as comprised of a number of such transverse wave movements. In ordinary light, however, such as that from a light bulb, these waves are not all vertical. Some are horizontal, and the rest are inclined at all angles across the direction of the ray. Figure 52 gives some idea of the vibrations of an ordinary light ray which we see head-on.

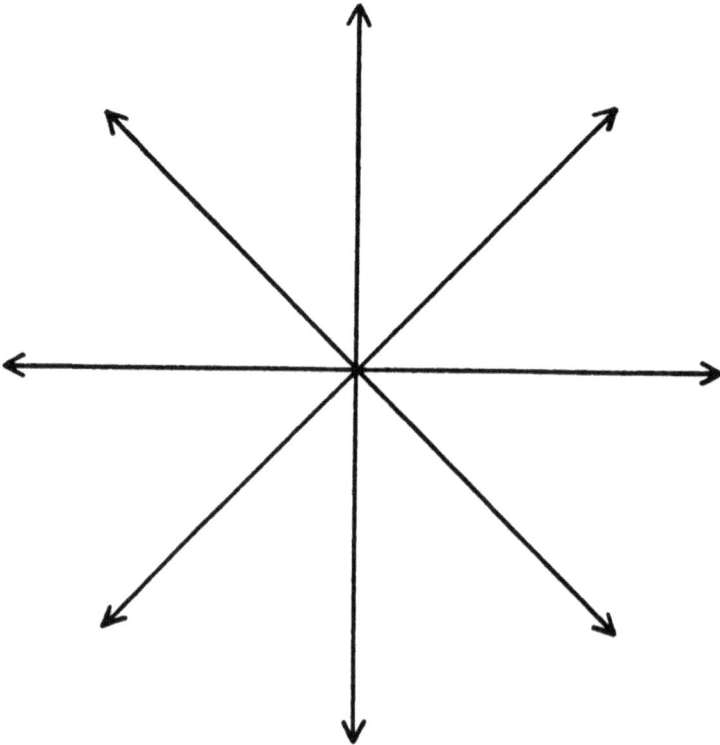

Figure 52. Schematic representation of a ray of non-polarized light, seen head-on.

In contrast to this sort of light, we can also have "plane-polar-ized" light. In the latter, the vibrations take place mostly in one direction, for instance vertically across the beam. This type of light is produced when rays are reflected by a mirror or by the surface of a pool of water. As a matter of fact, though, not all the light will then be polarized. The proportion varies according to the angle which the rays made with the reflecting surface. A consider-able proportion of the light which reaches us from the sun is also polarized. To be sure, before it reaches the earth's atmosphere it is not in that condition, but when the rays strike certain minute particles in the outer atmosphere, some of them are caused to scatter, and this fact is responsible not only for the blue color of the sky but also for the appearance of plane-polarized light. Both the proportion of the sun's rays which are thus polarized and the angle of polarization vary in different parts of the sky, according to a quite regular pattern. As the sun moves over the sky's hemi-sphere, the pattern moves with it. In other words, it is not neces-sary to see the sun itself directly. One can find out where it is from the polarization pattern of the sky. In order to do this, how-ever, we must first have a sure-fire way to determine or record that pattern. A great help to us humans, in this case, is a plastic polar-oid filter. Like other filters that fulfill the same purpose, it will let through only those light waves which are vibrating in one par-ticular plane. To get some idea of how such a polaroid filter works, we can think of it as being like a very narrow slit which will let a vibrating wave through only if slit and wave are parallel to each other. Therefore, a plane-polarized light ray, seen through one of these polaroid filters, will seem to be brightest when the filter is held in a very special position. If we turn the filter, the light intensity decreases, to become practically zero when we have turned through a right angle. When ordinary light is passed through a polaroid filter, all the waves are held back except those vibrating in the plane of the filter—or, in our comparison, in the plane of the "slit." Therefore, the light that comes through will be plane-polarized.

Honeybees, now, are definitely able to determine the light-polarization pattern of the sky. This is shown clearly by experi-ments in which a beehive was placed in a darkened shed. Some bees would still dance their figures-of-eight and indicate the direction of the food source on the alighting board, but they would only do this when a wide tube open at both ends was put through the roof of the shed with its lower end over the bees, so that the dancers could see a piece of blue sky through it. When a polaroid light filter was placed over the tube's upper end, the dancers could be forced to shift the direction of their figures through a certain angle by turning the filter through the same angle. We must assume that the numerous units of a bee's eye function somewhat like "polaroid analyzers," devices that can be made by arranging different triangular pieces of polaroid in a special way. It would lead us too far afield to explain this in detail. Also, there is no absolute certainty about this matter as yet. Many scientists are still working on the problem.

The ability to recognize polarized light and to find out *how* it is polarized is being found in more and more animals, although thus far they are all in the group of the hard-bodied animals or arthropods, which includes insects, scorpions, spiders, lobsters, crabs, and their allies. It seems that the eyes of horseshoe crabs, so abundant along the eastern seashores of the United States, are particularly favorable for studies concerning polarization. In the future, we shall certainly hear more about them. But let us return to the honey dance.

Russian scientists have found that an artificial dance effect can be obtained by feeding the honeybees in the hive a sugar solution mixed previously with, say, clover flowers. Von Frisch, who has repeated these experiments, found that in this fashion the visits to red clover in the field could be increased twenty-twofold! Moreover, the bees worked longer and more intensively. It is also possible to induce visits to odorless flowers such as those of the potato plant by spraying them with, for example, lavender extracts, feeding the bees at the same time a lavender-scented honey in the hive. Even though these experiments are not successful with all flowers, they certainly hold out a lot of promise in connection with the pollination of fruit trees, the increase of honey production, and so on.

The events that take place during the swarming of honeybees are fully as fascinating as the honey dances which we have just discussed. To understand them, let us briefly return to the life cycle of the bees in a hive.

Recall that, in the spring, a bee queen starts out by laying eggs which will eventually produce worker bees. Then she begins to deposit unfertilized eggs which will give rise to drones. After that, she will again produce fertilized eggs, destined to be queens. The larvae hatching from the last category get very special care from the worker bees and also are fed a special food, royal jelly. As a result, their development is very fast, requiring only from 16 to 17 days in comparison with 20 for the workers and 24 for the drones.

When one or more of the young queens have reached maturity and are about to leave their cells, the bees in the hive get into a state of frenzy. The most advanced of the queens, still concealed in her little prison, will produce a most peculiar, screeching sound. As soon as this is heard, whole crowds of young bees will begin to mill around in her vicinity, buzzing loudly. Other bees, however—mostly the older workers—leave this maddening scene. Some of them crawl out of the gate of the hive and back in again, as if to find out what the weather is like. If it is good, they will assemble around the old queen who, by this time, is just about ready to kill her rival with her sting. The return of so many of her subjects, however, somehow changes her mind. The upshot of it all is that the old queen leaves the hive with a large number of bees—5,000 to 30,000. They carry with them as much honey as they can hold, so that a good-sized swarm may well weigh 9 or 10 pounds.

In the hive, the young queen now emerges, after long screeching. Very rapidly, she moves to the cells of her sister queens. Sitting down on the lid of the first cell she reaches, she thrusts her poisonous sting through it a number of times, killing the hapless young animal inside. After that, it is the turn of the young queens in the next cells to be murdered. The throne of the new queen is indeed built on the dead bodies of her rivals! When her grisly work is completed, she will now fly out, surrounded by a whole crowd of those gay young blades, the drones. This is her bridal flight, or honeymoon. One of the drones will briefly become her husband, or prince consort, and after her return to the hive a few hours later the young queen is ready to begin fulfilling the heavy task of being the mother of her people. She will now start laying the eggs which will produce workers, drones, and new queens. When one of the last has reached the point at which *she* begins to screech in her cell, it is the reigning young queen's turn to bow out and leave the

hive with a swarm of her subjects. But sometimes this is not possible because of bad, rainy weather. In that case, the queen will
remain in the hive at the expense of the lives of her daughters, the
young queens, who are murdered.

In contrast to some of the worker bees, the drones never hibernate. They are real parasites, eating a lot without doing a lick of
work. As long as swarming continues, they are tolerated in the
hive. In July or August, however, when there is no further chance
that queens will emerge, the workers drive the useless drones out,
often killing most of them. Even those drones which have not
hatched from their cells do not escape the fury of the workers.

Let us now follow the further adventures of the bees that have
left the hive in a swarm. Normally, they are scooped up by the
alert beekeeper, who will provide them with a new home, and there
will be no real problems. What happens, however, to bees that are
not too closely supervised or to honeybees still living in a more or
less wild state? This is a very perilous period in their existence as
a colony! A good new home has to be found, the sooner the better,
and complete unity of action is necessary lest all perish. The following description of the wonderful things that now take place is
based largely on the work of M. Lindauer, originally a collaborator
of von Frisch. The events are so strongly reminiscent of a political
convention in the United States that it is very hard to avoid
humanizing the bees. With tongue in cheek, we shall simply follow
Lindauer.

It is clear that if literally all the bees in the swarm had a say in
the matter of selecting the new home site, complete chaos would
result. Therefore it is a good thing that the decision is left entirely
in the hands of a small "committee," the scout bees. Of course
there is no such thing as a scout "profession" among the bees;
there just could not be, because many generations may pass without swarming, so that "practice" is entirely out of the question.
The scouts simply are foragers who found little left to do in the
hive when the period of swarming approached. All the supply
chambers in the combs were full of pollen and nectar, and the
honey sacs of the honeybees were, likewise, brimful, so that the
foragers were no longer able to deliver their nectar to other bees.
In this unusual situation some of them, rather than remaining idle
in the hive, seized the initiative and started home hunting, even
before swarming really took place!

Once the swarm has been formed, the scouts seem to intensify their efforts. They fly off in all directions, and when they have found a suitable place, such as a hollow tree, a hole in an old wall, or a chimney, they come back and tell their colleagues about it in dance language. But what is a suitable place to them? We can find this out by presenting the bees with a number of artificial nest sites not too far from the swarm, such as wooden boxes in various sizes, baskets, and so forth. We wait and see which of these they choose. It turns out that *size* is something very important. Preferably, the new home should not be much bigger or smaller than an artificial beehive, but there are exceptions to this. Thus, bees from a very small swarm will choose a small home, and a big box becomes acceptable to them only after we have first padded it on the inside to make it smaller. The construction of the new home is important, too. A wooden box is considered better than a basket, but a hole in the ground is even better. It seems that the bees want to avoid draft at all cost. Protection against strong wind is also crucial; a box covered with twigs, put under a tree, is much more attractive than just a box. But the direction of the wind may change from one day to the next, or even in the course of one single day. Therefore, it is a good thing that the scouts visit each of the likely spots several days in a row and several times each day. They will only advertise a spot strongly if it is good on all occasions.

It is very interesting that distance counts so heavily; faraway spots are definitely favored over spots close by! A moment's reflection will, of course, make the significance of this clear to us. The farther away the swarm moves from the mother colony, the less severe will be the future competition for food, that is, flowers, with the inhabitants of the old hive.

It happens, sometimes, that a number of scouts return to the swarm at the same time, and as many as twenty different spots may be announced simultaneously. The rules which the bees follow in their danced announcements are very much like those observed by foragers returning from a rich food source, the main difference being that this time they have such things as brick dust or soot on their bodies instead of pollen and the smell of nectar. Both distance and direction, then, are indicated by the returning scouts. However, since so many different spots are involved, one can easily see that a mad medley of dances results. How in the world, one may ask, will the bees ever reach agreement? The manner in which they do it is certainly well calculated to put us humans to shame; I wish I could say that our political conventions were decided as simply and elegantly.

What is essential, really, in the first place, is that each scout bee be perfectly honest in her reporting. She should be quite enthusiastic about a good spot and not so excited about one that is only fair. This is indeed the case, for a scout will announce an excellent site by a truly vivacious dance which lasts for an hour or even longer. On the other hand, a home that is acceptable and no more is indicated by a lazy dance of only a few seconds' duration.

Secondly, a scout should not be pigheaded and should not stick to her first choice when other bees have better possibilities to offer. Again, this is the case. Some scouts are "converted" by the enthusiasm which a certain scout may show. They "join the band wagon," so to speak, and will visit the spot indicated by their colleague. Once they are convinced that her spot is indeed better than the spots which they had discovered, they will now begin to campaign for her, and will dance in the way she dances.

But let us illustrate all this by an example which I have borrowed from Lindauer. In the afternoon of June 26, a scout bee from swarm E ("Eckschwarm") came back for the first time with information concerning the spot S, which was finally chosen by the whole swarm. That afternoon, only one other nest site was announced by another scout. In the course of the next few days, however, a total of 27 different spots were considered in the deliberations that took place. In the morning of June 29, there was dancing by 22 scouts indicating 11 different sites; 7 of the scouts had a preference for spot S. Late in the afternoon, many more had joined the band wagon, so that 61 scouts were now in favor of S, but 2 stubborn ones still held out for another direction. Even the next morning, the haggling was not quite over. At 10 o'clock, however, agreement was complete, and the swarm took off to the new home, S.

Sometimes the decision is not so easily arrived at, especially when at last there are only two factions, about equal in numbers, among the scouts. It is only in very exceptional cases, however, that no agreement at all can be reached. In our climate, this usually means that the swarm perishes.

# 13

# *Ambushes and Traps*

In many ways, it is a good thing that this earth of ours is "shrinking" constantly. In the old days, it was sometimes necessary to organize a complete expedition if you happened to be interested in certain plants; now, you just order them somewhere and grow them in your own garden or greenhouse. This gives me the courage to present here a European plant which I consider one of my special friends; the spotted arum lily, *Arum maculatum,* which in summer shows its bright-red "berries" so beautifully in shady places. In the southern part of the Netherlands I have seen the wild plants by the dozens, and I am sure one can also encounter them without any trouble in the vicinity of Oxford, England. Usually, twenty or thirty of the red berries are grouped closely together on a thick green stalk which may be a foot or so high. In southern Europe, there grows another but closely related species, *A. italicum,* with very attractive, light-veined leaves. In the parks of Florence, it is as common as a weed.

The spotted arum usually flowers in May. What we call a "flower" here is not really a flower but a complete inflorescence, surrounded by a big protective leaf, or enveloping bract, the spathe. Since we do not want to bore you with long descriptions, we suggest that you simply look at Figure 53 to get an idea of the way in which the separate male and female flowers in such an inflorescence are arranged. The bristles which can also be seen are modified, sterile flowers. We shall soon see that they play a very important role.

As the spathe unfolds in the afternoon of the first flowering day, the club-shaped, naked appendix which sticks out begins to develop a very unpleasant odor, somewhat like a mixture of carrion and urine. At this stage of the game, the female flowers at the base of the central column are already mature. Each has, for a stigma, a whitish brush completely wet with a sweet, slimy fluid. When you open the floral chamber of your arum lily, a little later, you may be in for a big surprise: a terrific number of small gnats will be liberated, as many as 4,000 sometimes, from one chamber! It is obvious that they were trapped there, but how and why did they get in, in the first place? Experiments have shown that it is the carrion smell of the appendix that attracts them. When one makes a glass imitation of our arum lily, it is possible to accumulate in the chamber the same types of insects which are trapped in the natural flowers, provided the model contains a cut-off, fresh spadix, or, otherwise, some decaying blood. (It is of course wise to mix the blood with some glycerin so that it will not dry out too quickly.)

I do not have to point out that carrion is something which is very much in demand in the animal world; blowflies and various types of beetles are very fond of it, too. However, most of these fellows are much bigger than the little flies and beetles which we saw in the chamber, and the reason why we do not find them trapped is that the arum flower discriminates against them, using the brush above the male flowers as a sieve or collander, so to speak, which allows only the smallest gnats and beetles to pass. But—someone will say—if it is the appendix that attracts insects, why don't they stay in the upper, exposed part of the arum lily? Fritz Knoll, who has done beautiful work on this whole affair, has shown that it is all a matter of slipperiness. Obviously, the insects must have *fallen* in, sliding down along the inner surface of the spathe. At first sight, you may find this very hard to believe, for we know that many insects are past masters when it comes to the matter of hanging on to difficult surfaces. On a somewhat rough surface, they use their little claws, but when the surface happens to be very smooth they use the tiny suction disks with which their feet are also equipped. So how can one beat them?

The trick with which the arum flower has done it is to provide the inner surface of the spathe with countless tiny oil droplets. Now, the little insects are helpless. If they cannot fly away—and that is what the bigger flies and beetles do, of course—they slide down relentlessly, through the brushes above the male flowers, into the chamber. It stands to reason that they try to escape from this, but they soon discover that an ascent along the inner chamber walls is utterly impossible; these are just as bad and slippery as the upper spathe surfaces. Trying the alternate path upward across the female flowers, they find themselves blocked by the lower barricade of brushes shown in Figure 53. So they have to stay. This is not as bad as it seems, however, for on their trips back and forth across the zone of the female flowers the insects have come in contact with the slimy, sweet sap of the stigmas, which they like very much for food. If they happen to have some arum pollen on their bodies, part of that will be left behind on the stigmas, where it soon begins to germinate.

During the hours of darkness, the insects which have had so much opportunity to gorge themselves will usually quiet down. At this time, however, the stamens of the male flowers will gradually open, and the insects in the chamber will be bombarded with a steady rain of pollen grains from above. The next day, therefore, they will find themselves "tarred and feathered," so to speak. Resuming their efforts to escape, they discover that the chamber walls are just as bad as before, but the brushlike barrier organs have lost their stiffness. They have wilted a little and now expose a wrinkled surface, so that the insect claws find them to have a lot more "grip." Thus, the gnats can crawl upward, leave the female flowers behind, cross the zone of the male flowers and the uppermost region of "brushes," which have also wilted, to find themselves, at long last, in the open. The appendix now has lost its carrion smell, and the prisoners fly off. However, their responses have not changed. As soon as they notice the smell of another arum flower which is in the "female" stage, they will be attracted and in a short while will again be effecting pollination in another prison chamber.

The flowers of *A. maculatum* and *A. italicum* show still another peculiarity, which they share with many of their relatives: at the time when the events which I have just described are taking place, the temperature rises rapidly. One can really feel that the flowers, and especially their appendices, are warmer than their environment. You can also show it with the simple thermometer experiment illustrated by Figure 54. A difference of 15° to 20° C (about 27° to 36° F) was early observed by Federico Delpino, who studied these flowers around the middle of the previous century. He sug-

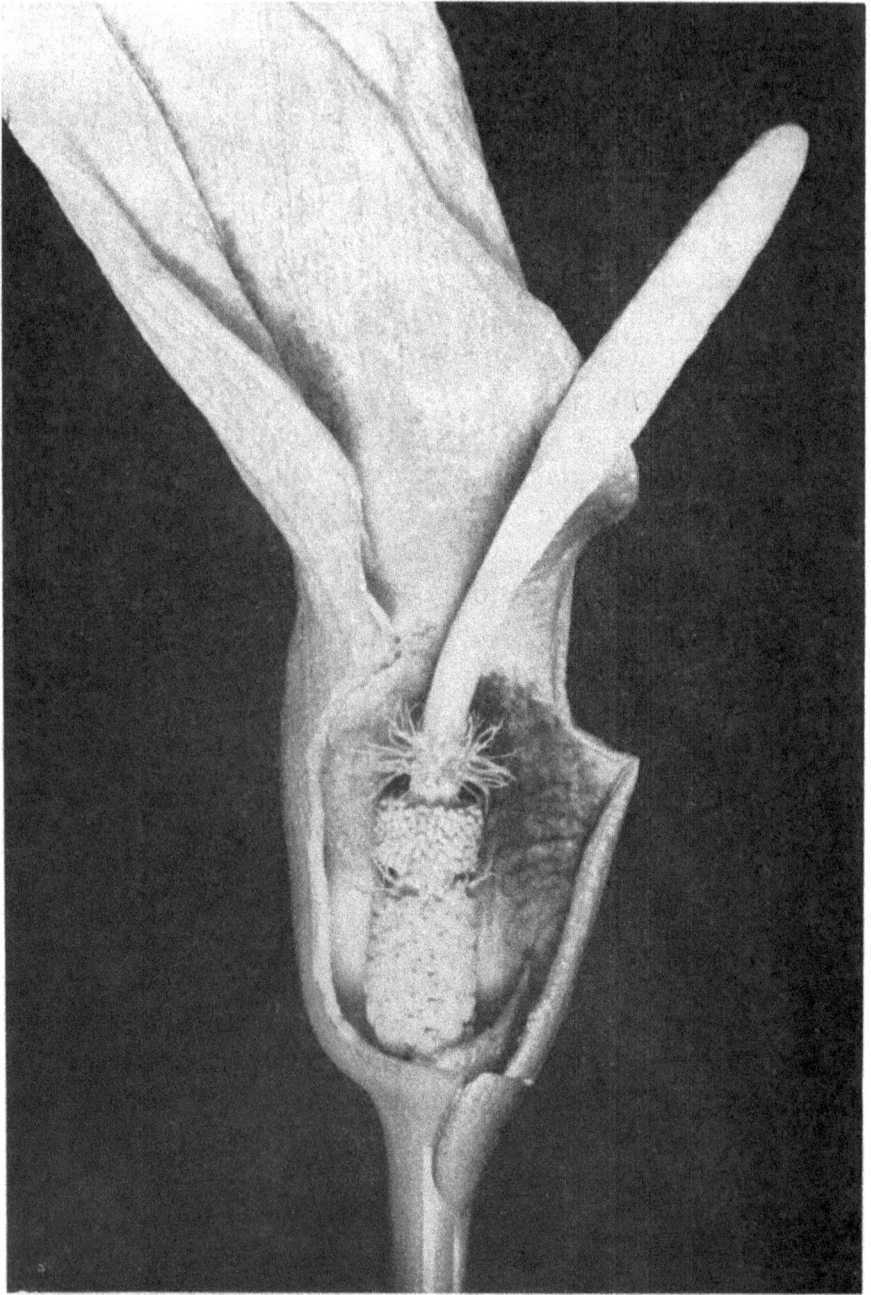

Figure 53. An inflorescence of *Arum maculatum*, a wild arum lily from western Europe, cut open. The female flowers are at the base, followed by a group of bristles (sterile flowers). Above that, we see a group of smaller, male, flowers and another group of bristles. To attract pollinating flies, the naked upper part of the central column, the so-called "appendix," produces a carrion smell and also the high temperature necessary to spread this odor.

gested that perhaps the high temperature helped to attract the insects, but this idea turned out to be wrong. If small electric heaters are put in the artificial flowers instead of rotting blood, no insects are attracted. The main, or perhaps the sole, function of the high temperature must, therefore, be the volatilization of the smelly substances produced by the appendix.

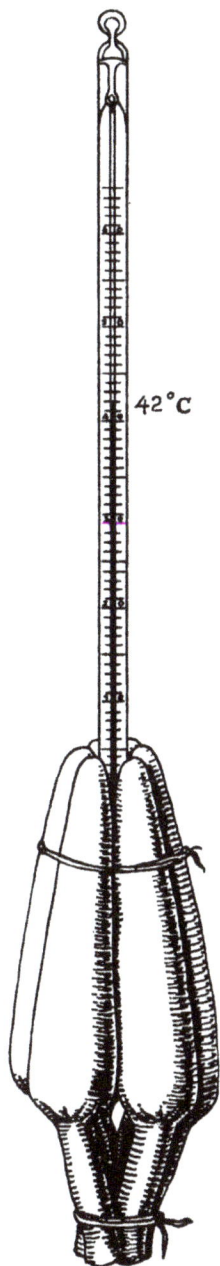

42°C

In the southeastern part of the United States we find the various pipe vines, for instance the famous Dutchman's pipe, *Aristolochia durior*. The real Dutchman's pipe from western Europe (*A. sipho*) is often grown in medicinal gardens in America. The name gives a clue to the shape of the flower, which essentially is a slightly curved tube with the upper part flanged out to give a "flag" and the lower part inflated to form a chamber (Fig. 55). As we shall see, these flowers operate like well-constructed fish traps. They are placed in the axils of the upper leaves. At the time when they open, they have an almost vertical, erect position. Their colors are not very conspicuous but, on the other hand, they have a peculiar smell that attracts small gnats (*Chironomid* flies, especially *Ceratopogon*). When these insects try to sit down on the flag, they slide down immediately. To be sure, we do not find oil droplets here, as we do in the case of *A. maculatum*, but there are tiny granules of wax that come off easily and fulfill the same purpose. Furthermore, the narrow flower tube has a number of hairs which point downward and which are attached to the wall of the tube in a very ingenious way. On a rather small "joint" cell, which forms the connection between hair and tube surface, we find the big, roundish basal cell of the hair, which is attached *excentrically*

Figure. 54. Development of heat by flowering Arum lilies. The sterile upper parts of a half dozen inflorescences of *Arum maculatum* have been cut off and tied around the mercury reservoir of a simple thermometer.

(Fig. 56). Thus, the insects have no trouble in going down, but a return is made impossible because, as soon as the hair is pushed upward, the thick basal cell will come into contact with the tube wall, making further movement in that direction impossible.

Before long, the gnats find themselves in the chamber, at the bottom of which is placed the peculiar structure known as the

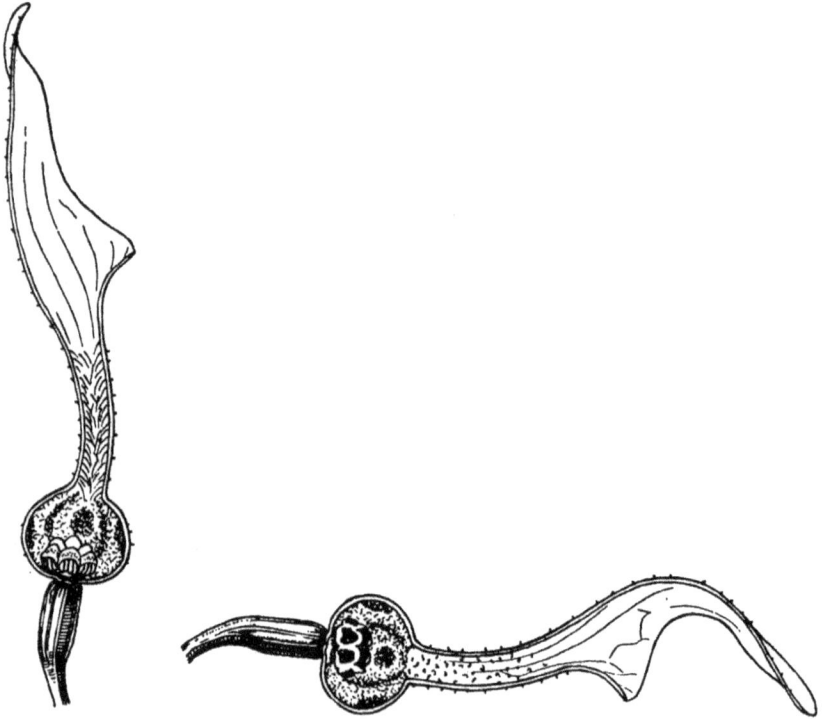

Figure 55. Two flowering stages in Dutchman's pipe (*Aristolochia sipho*). Left: flower erect, hairs on the inner wall of the flower tube stiff and pointing downward, making escape of the trapped small flies impossible. Right: flower limp, hairs wilted: the prisoners can escape.

gynostemium, which is formed by fusion of the six anthers with the pistil. At this stage, the six stigmas are ready to receive the pollen grains which the insects possibly brought in. The anthers, however, are still closed. For two to three days, the gnats have to remain in the chamber, feeding on nectar. Gradually, the anthers open and the visitors are powdered with pollen. During this time, the flower bends over, so that it now occupies an almost horizontal position. The hairs in the tube wilt, and so does the flag, and the gnats escape, still covered with pollen which in another *Aristolochia* flower will effect fertilization.

The same general flower structure and pollination type can be observed in many other tropical and subtropical *Aristolochias*, some of them with very big flowers such as the "pelican flower," *A. grandiflora*. Occasionally, you can see these in greenhouses, and there is one—a very rare one, unfortunately—for which I would like to ask your special attention: *A. lindneri*. The flowers of this plant, which occurs in the wild state in northeastern Bolivia, combine in a marvelous way the "slide trap" principle with which we

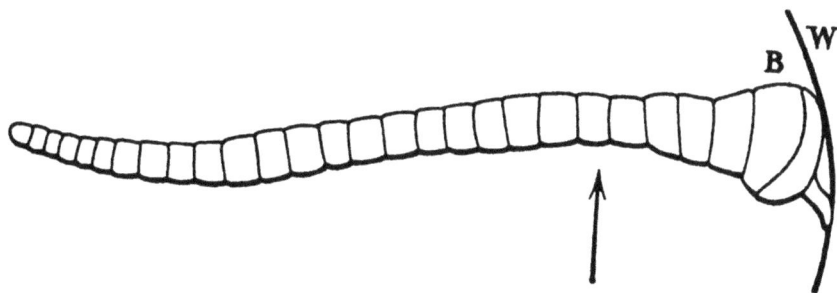

Figure 56. Hair from the inner surface of the flower tube of Dutchman's pipe, *Aristolochia sipho*. The arrow indicates the direction in which the small flies trapped in the floral chamber try to escape. Notice the peculiar construction of the hair, with basal cells B being pushed against the wall of the flower tube W, preventing further movement.

are now familiar, with certain light effects. Instead of the one "flag" which we found in *A. sipho*, we have here rather big upper and lower lips. The tube, which is somewhat curved, is similar to that of our Dutchman's pipe in that it has slippery walls and a number of hairs pointing downward; however, it is quite dark. Between the tube and the chamber we find a "diaphragm," a crosswall with a small, funnel-shaped opening in it. In contrast to the tube, the chamber is light and transparent, with the exception of a dark nectarium which is placed on the ceiling of the chamber and a dark, ring-shaped zone around pistil and stamens (Fig. 57). Between the dark ring and the gynostemium we find another, very light, zone. Insects, attracted to these flowers by a very strong feces-like smell, will sit down on the lips which are spread out almost horizontally. Walking back and forth, there is a good chance that they will fall down into the tube which is more or less vertical.

Here, they find themselves in a tomblike darkness. However, one light spot beckons them, namely, the hole in the diaphragm. Moving toward it, they get into the chamber. There, the spot that attracts them most is the light zone immediately around the gyno-

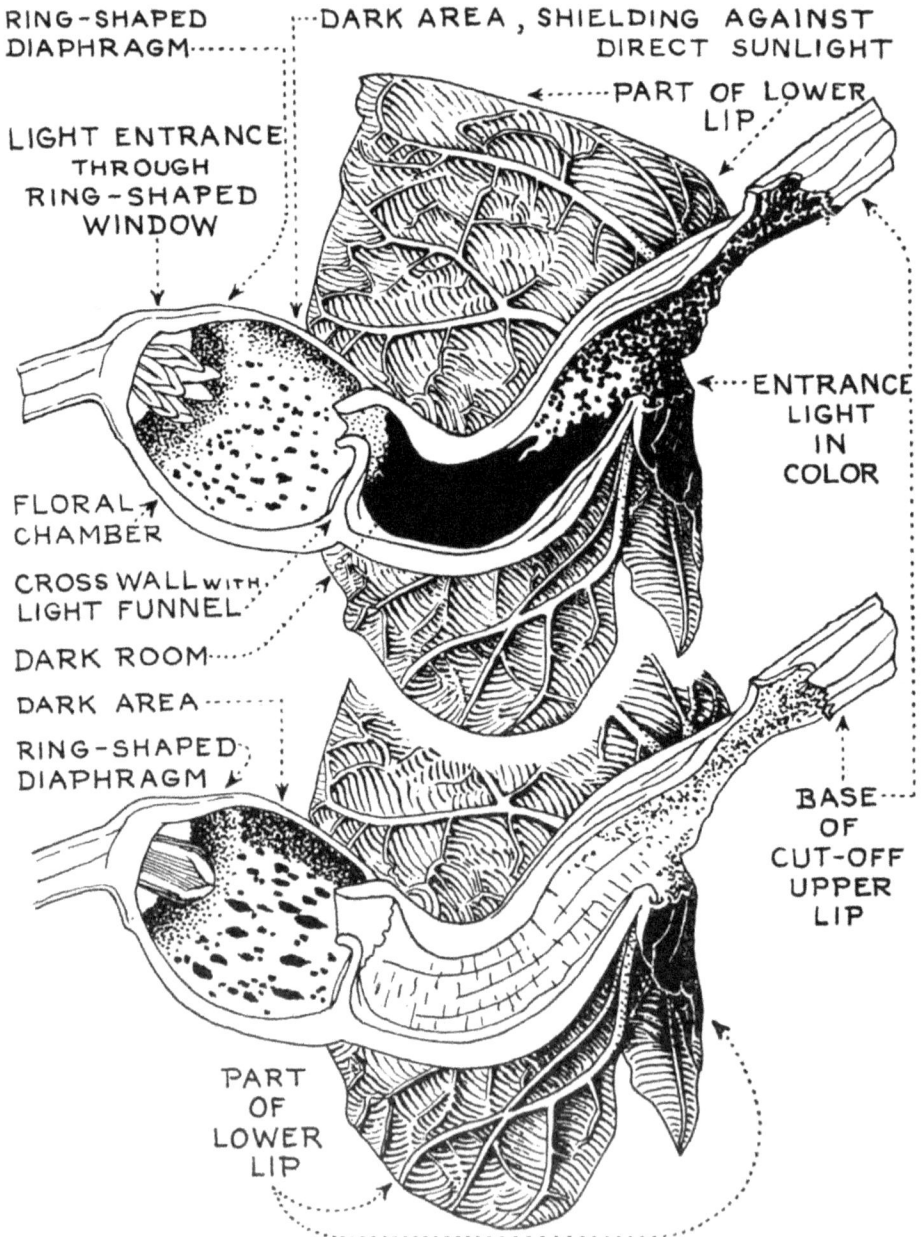

Figure 57. Two stages in the flowering sequence of *Aristolochia lindneri*, one of the most sophisticated trap flowers known to man.

stemium; consequently, there is a pretty good chance that they will deposit on the waiting stigmas the pollen which they brought with them from the outer world.

On the second day of their imprisonment, a whole series of changes takes place: the anthers shed their pollen, the hole in the diaphragm becomes larger, the tube loses its dark color and affords a passage, and the lips wilt. So, covered with pollen grains, the prisoners reach freedom again.

Talking about ambushes, it is impossible to skip the flowers of the beautiful lady's-slipper orchids (*Cypripedium* species). Some people compare them with wooden shoes. I am not sure that all of them are equally well suited to act as traps, but there are at least several in America that will do a good job; for instance, the moccasin flower of the East and the beautiful *C. montanum* of Washington and Oregon.

The slipper part is extremely smooth on both inside and outside. There is only one part on the inside of the shoe that forms an exception to this, namely the "heel," where we find a strip of tightly packed, long hairs (Fig. 58). Insects such as bees and flies can get into the "wooden shoe" with great ease through a hole in the upper part. Once inside, however, they are unable to climb the smooth and steep walls unless they move along the hairy strip which will take them to the gynostemium, that is, the column formed by the fused pistil and stamens. It seems that, in doing this, they are also guided by light effects, for certain parts of the wall of the wooden shoe are transparent, somewhat like glass windows. When the insects finally escape through the hole over which the gynostemium is placed, they just cannot avoid hitting, first, the surface of the stigma and then one of the two anthers. Very significantly, the pollen grains are in this case connected with one another through a thick, oily fluid, so that they form one sticky mass which will attach itself to the visitor without any trouble. So, the insect will not go to the stigma of the next lady's-slipper "empty-handed." It is interesting to note that, in most other orchids, the pollen grains are combined in compact packages, pollinia, which as a rule are not particularly sticky.

Before we turn our attention to other matters, let us briefly consider the various members of the milkweed family, the Asclepiadaceae, to which our butterfly weed (*Asclepias tuberosa*) and the common milkweed (*A. syriaca*) belong. The milkweeds certainly are interesting in a great many ways. Here, we can only discuss their flowers.

In Asclepiadaceae, the five anthers are fused into a ring around the pistil, so that a central column with a flat "summit" is formed

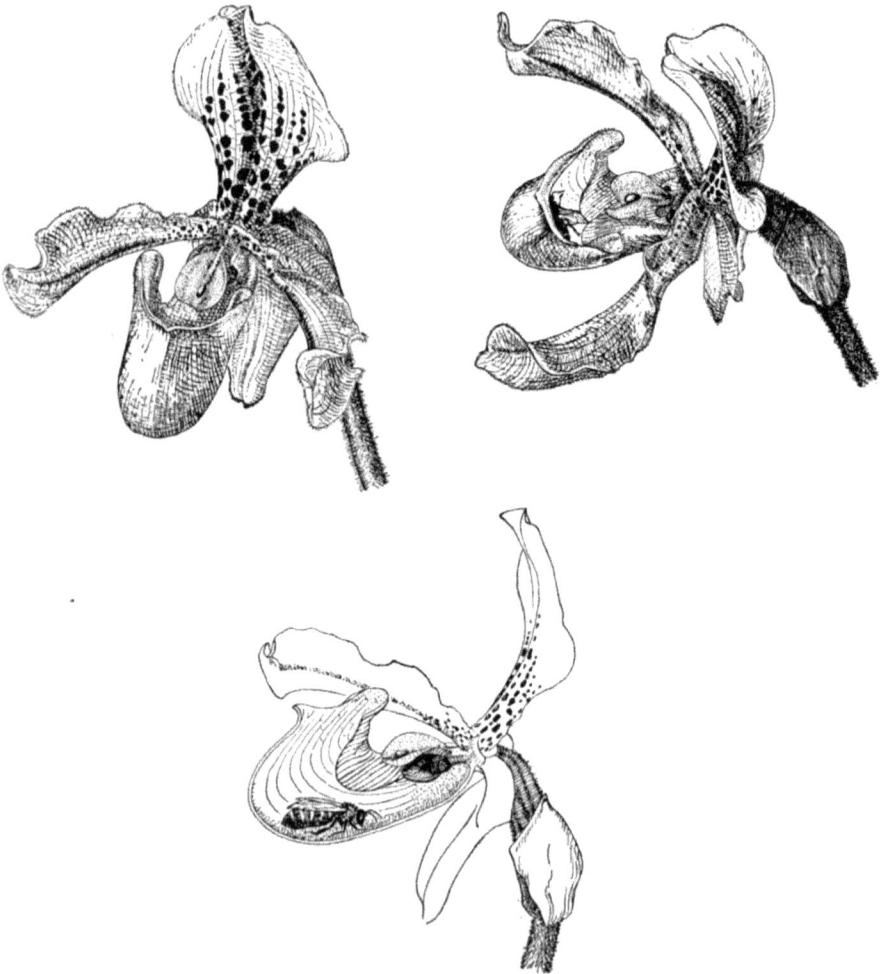

Figure 58. Lady's-slipper orchid (*Cypripedium*), an example of a trap flower.

in the flower. With a little bit of trouble, however, we can still distinguish the individual anthers. For one thing, each has on its back a beautiful nectarium, which reminds one at the same time of a devil's horn and a honey cup. One has to see them to believe this (Fig. 28, page 79). Furthermore, the stamens are separated by very narrow slits running vertically along the column. At the top of each slit there is a peculiar tiny structure, the translator, visible as a black dot. Examination with a good hand lens reveals that this dot actually is a little clasp, a horny plate which is folded lengthwise so that the edges almost touch each other. The idea is that an insect will alight on the flat summit of the central column and that one of its legs will slide down in one of the vertical slits. If so, it begins to struggle to free its trapped leg and, if it is strong enough, will pull the column up vertically. Inevitably, the leg then will follow the slit, which narrows considerably in the upward direction, and will come into contact with the translator, which clasps itself firmly around it. By means of strong thin little "stems," each translator is connected with two waxy pollen masses or pollinia, one from each of two neighboring anthers. Still pulling, the insect will thus draw out a pair of minute "saddlebags" (see Fig. 34, page 87). If it succeeds in flying away, the stems of the pollinia will, after a while, begin to curve, so that the pollen packages now face downward. There is thus a very good chance that, during the insect's visit to a following milkweed flower, they will stick to the latter's stigma. When the insect takes off from there, the stems break and pollination has been effected. However, it is not at all unusual to find a number of dead flies in the milkweed inflorescences, or even bees (see Fig. 35, page 88). They were not strong enough to liberate themselves and remained trapped.

Ceropegias (also members of the milkweed family) have flowers very similar to those of Dutchman's pipe, so that here we have a combination of the fish-trap and the bear-trap or clasp idea. The South African stapelias, starfish flowers (Fig. 59), victimize flies in still another way. Their blooms are, in general, characterized by a smell and a color very much like that of rotting meat. They will therefore attract carrion flies, which will take care of pollination. The female flies are fooled so completely that they will deposit their eggs (or maggots, as the case may be) in the *Stapelia* flowers. Finding no food, the larvae must perish. It is therefore safe to say that *Stapelia* behaves as a parasite toward the flies. Paradoxically, if there should ever be sufficiently large numbers of *Stapelia* in a given area, they might very well dig their own grave by killing off all their potential, future pollinators, which they do by prevent-

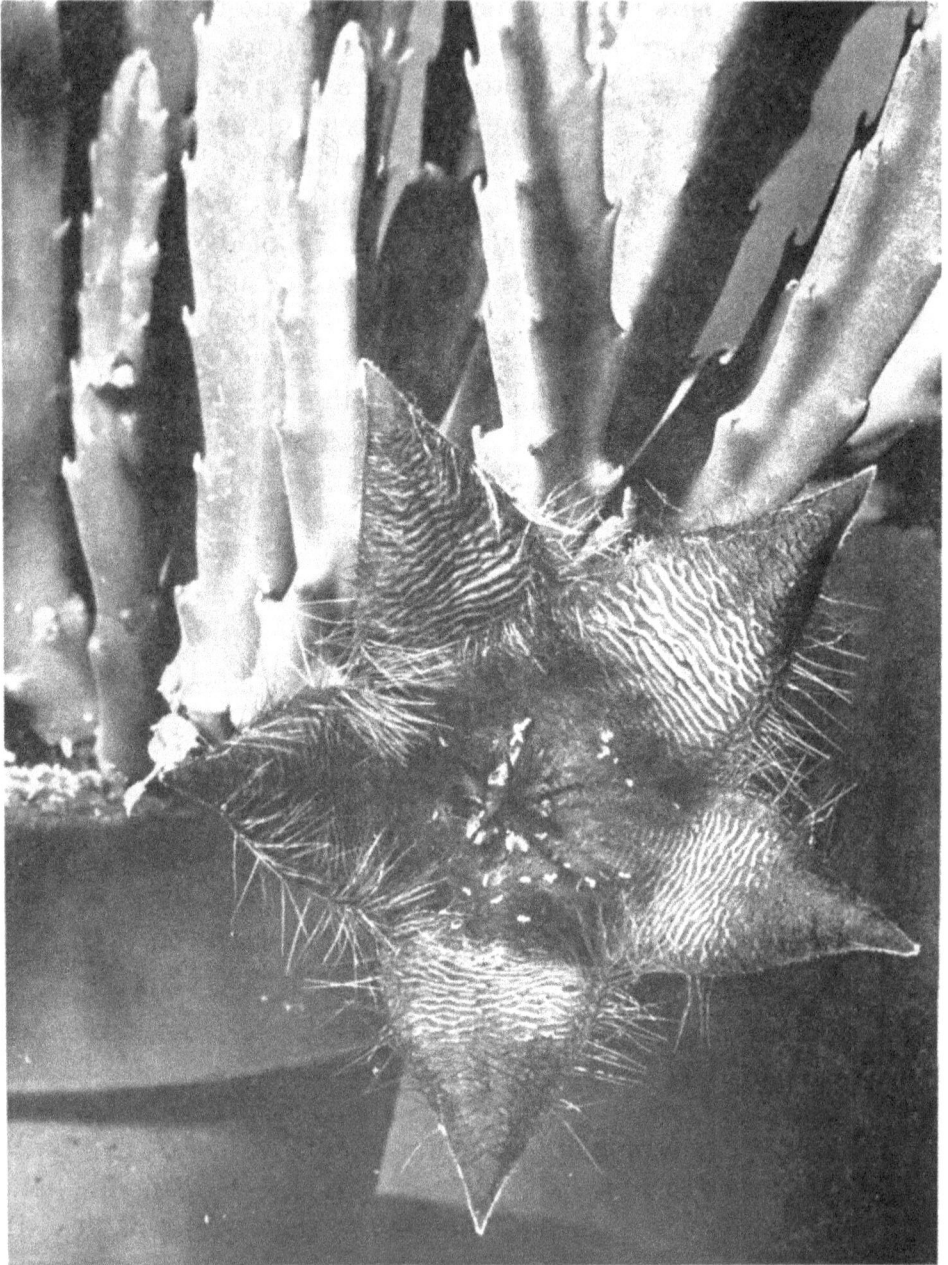

Figure 59. A starfish-flower (*Stapelia*). Flies are attracted to it because it has the smell, and also the general color, of rotting meat. The animals are fooled to such an extent that they will deposit their eggs in the flower. The maggots (white specks in our photograph) perish.

ing the flies from producing offspring. Fortunately for the plant, this is never the case, stapelias being desert-inhabiting forms with a low population density; there will always be enough flies left.

Reminiscent of starfish flowers, in that they also imitate rotting meat by color and smell, are the gigantic flowers of *Rafflesia* (Fig. 60), an Indonesian plant which, being a root parasite, is largely subterranean.

After all these woeful tales of trapped flies, I could hardly blame you if you came to the conclusion that Sprengel was right when he indignantly spoke of the "dumb" flies, fit only to be fooled and incarcerated. It is indeed interesting to note that, in contrast, flowers enlist the cooperation of the "smart" bees by offering them nice colors and attractive smells and sweet nectar.

And yet, in all fairness, we must point out that flies are by no means the only insects victimized by milkweed flowers. Even big fellows such as bumblebees and carpenter bees will at least be bothered, especially (of course) when the flowers happen to be large. I remember that in the East Indies we had big blooms on our biduri plants (*Calotropis gigantea*), 10-foot-high shrubs of the milkweed family which we sometimes cultivated for the sake of the beautiful tufts of "silk" on the seeds. The only insects we ever saw on the flowers were the sturdy carpenter bees, which on the island of Java seem to play the same role that bumblebees have in Europe and North America. Although, in general, these carpenters like to go about their business in an unhurried way, so that they may spend quite a bit of time on one flower, they would, in the case of the biduri, always fly away after a very brief visit, describing a wide circle in the air as if they were scared. It is true that after a while they would come back, bribed by the abundant sweet nectar. Even so, it was quite clear that the pollinia attached to their legs bothered them, for the bees would often energetically try to rub them off on twigs. Fortunately from the point of view of the plants, this last operation never was quite successful, as evidenced by the fact that we never noticed a lack of fruit and good seed. The pollination system here must indeed be foolproof.

Very peculiar pollination problems are sometimes created when plants of the milkweed family are introduced into another country. This has happened with *Arauja sericifera*, which in its native Argentina and southern Brazil probably is pollinated by large bees,

Figure 60. Flower of a *Rafflesia* from Sumatra. Possessing a carrion smell, it is probably pollinated by flies.

just like our biduri. In California it attracts hawkmoths, and in many cases these animals are not strong enough to free themselves once they have been caught. In the morning, the unfortunate victims can be found, each dangling from a flower by its proboscis. They can only perish miserably, hence the name "cruel plant" for this *Arauja*. H. Cammerloher, who has written an excellent little book on pollination, tells a similar sad story about an exotic, large-flowered *Periploca* which he found growing in a garden in Rovigno on the Adriatic. This plant attracted honeybees in large numbers but killed hundreds of them. It also caught one small hawkmoth, *Macroglossa*, a member of the same species Knoll had used so successfully in his honey-guide experiments. Here, too, the animal had obviously lacked the strength to free its proboscis from the slit and had died its weird death.

However, enough of these gloomy stories! Let me wind up by presenting some unadulterated North American slapstick, with a happy ending to boot: the case of the calopogons or grass pinks, beautiful wild orchids of the eastern United States and Canada. They are usually found in peat bogs where the environment is quite acid, but with a little bit of care one can grow them very well in normal garden soil—on condition that mice and chipmunks are kept away, for these consider orchid corms an exquisite treat.

The "pretty calopogon," *C. pulchellus,* which is found as far west as Minnesota and Oklahoma and as far south as Cuba and the Bahamas, is indeed so attractive that even in faraway England and Holland people have been cultivating it since the middle of the eighteenth century. In contrast to what we see in most other orchids, the lip of the flower here is uppermost in the perianth, occupying an almost vertical position: a regular flag (Fig. 61). In its middle, we notice a low, lengthwise ridge which is densely covered with club-shaped, glandular hairs, thus providing a hold-fast and perhaps even some food for insects. (This structure, by the way, is responsible for the name *Calopogon,* or "beautiful beard.") The base of the lip, forming the connection with the rest of the flower, is so narrow and thin that it acts as a hinge. As soon as an insect tries to sit down on the lip's front surface, clinging to the hairs, the whole blade or "flag" quickly falls forward and down, onto the protruding column formed by the stigma and the lone stamen. Even the wind will sometimes do the trick, but the best "releasers" of the mechanism are two bee species belonging to the genus *Augochlora;* these seem to have just the right size, shape, and weight. Indeed, before the hapless *Augochlora* can say "Jack Robinson," she finds herself sprawling flat on her back on the

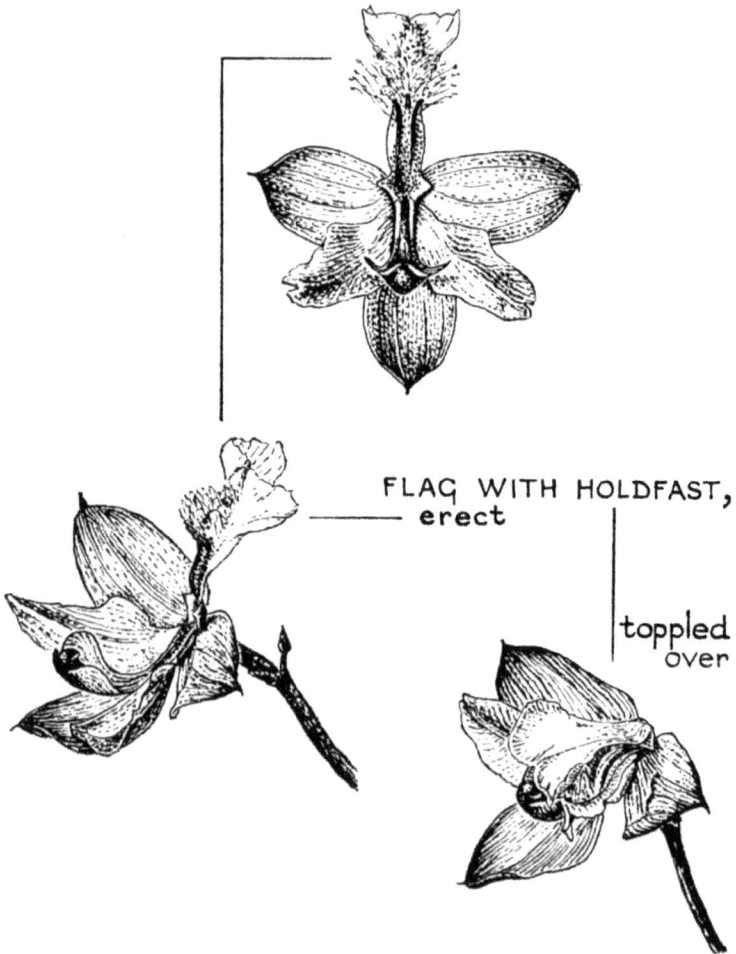

FLAG WITH HOLDFAST,
erect

toppled
over

Figure 61. Flower of a grass pink, *Calopogon pulchellus,* in various conditions.

upper surface of the column, which has a shape very much like that of the kiddie slides in our amusement parks, with ridges for margins. So nicely do these ridges fit around the insect's body that the animal, perforce, can only continue to follow this skid road downward. First of all, now, a sticky material from the surface of the column will attach itself to the back of the bee. Then, after having passed the stigma, the animal will hit the anther, which is placed just a little way downward. As a result, the stems of the four pollen masses, or pollinia, will pop out and attach themselves to the bee's sticky body—always in the same spot, on the first ring of the abdomen. Since the bee is still tumbling down, the main bodies of the pollinia must follow suit. Later, when the animal visits a second flower, the story is repeated. This time, however, the pollinia which the bee carries are left on the stigma during the ride down.

# 14

# *The Imitators*

There is more than one reason why it pays to have a bird bath or little pool in your garden. Even though it may only be a few inches deep and may be half full of mud and decaying leaves, you will be amazed by the teeming animal life in it. I would be willing to bet that one of your most regular customers will be the larva of the dronefly, *Eristalis* (or *Eristalomyia*) *taenax*. About 200 years ago, R. A. F. Réaumur—the scientist who is best remembered for his thermometer but who did some splendid biological work as well—beautifully described this "rat-tailed maggot." I think it had a special appeal for Réaumur, the physicist, precisely because of its peculiar tail. That organ is built like a telescope, with tube-like compartments that can be shoved in and out. The whole tail can thus be lengthened and shortened at will, and the larva can breathe pure, atmospheric air through it while feeding on decaying matter some 4 or 5 inches below the surface (Fig. 62).

Figure 62. Larva of the dronefly or "blind bee," *Eristalomyia taenax*. Often dwelling in foul, oxygen-poor fluids, it survives, thanks to a system of extension tubes that reaches all the way to the surface. The adult insect, a common pollinator on flowers of the "open" type, mimics the honeybee.

Our maggot, then, is really very much like a modern submarine equipped with a snorkel. Often this arrangement comes in very handy indeed; we must keep in mind that our little "rat" is frequently found in the most unbelievably foul, oxygen-poor environments. For instance, it just thrives in that heady mixture of cow manure and urine which is so highly prized by farmers, and in the fluid which accumulates in decaying carcasses.

When the larva is full grown, it leaves its delicious soup or its bird bath and pupates in the larval skin. The adult insect which emerges after some time, the dronefly, is just as fascinating as the larva. The "drone" in its name must mean male bee, for it looks almost exactly like a honeybee. The Dutch have a funny way of expressing this; they call the animal a "blind bee." (The word "blind" here simply means something that is not real. Anglo-Saxons use the word in the same sense when they speak of a passage that does not lead anywhere, as a blind alley.

Even the flight of the imitator (the dronefly) and the model (the bee) is somewhat the same, for *Eristalis* has rather long and stout hind legs which it usually keeps just a little stretched. Since it also lets them drag a little in flight, one gets the impression of a bee with pollen baskets. When ancient writers such as Ovid and Virgil claimed that bees originate in the carcasses of wild animals, we must conclude that they simply were early victims of our deceitful *Eristalis!* Even birds are taken in by the resemblance, as has been shown very nicely by G. Mostler, but I will discuss that later on.

Originally, *E. taenax* was an Old World species, but nowadays it occurs all over the globe, practically all the year round and sometimes in great abundance. Since its mouth parts are rather short, one will encounter it mostly on "open-type flowers," where nectar and pollen are easily accessible. The inflorescences of Umbelliferae such as Queen Anne's lace and angelica, and the flowers of ivy with their numerous little sweet drops are just about tops. Of course, other flies with short mouth parts will keep *Eristalis* company there. Some of these, such as the bluebottle flies (*Calliphora*), the metallic-green lucilias, and the hairy, parasitic Tachinid flies, are jacks-of-all-trades, and may also be found on meat, excrements, and carrion.

The closest relatives of *Eristalis*, however, the Syrphid flies, are such pronounced flower specialists that they are often simply referred to as "flower flies." Some of them are very much like the dronefly in their breeding habits, and you should, therefore, not be too amazed when some of the larvae in your bird bath produce colorful, wasplike, yellow-and-black flies instead of the rather plain, brown *E. taenax*. I do not want to bother you with their Latin names, but you may want to look out for a handsome, big fly with a death's head on its thorax, for another one which is somewhat tiger-like because it has lengthwise black and yellow stripes, and finally for a very close relative of *E. taenax* which looks almost exactly like a small bumblebee, dressed as it is in an orange-and-white coat of short fur. This last creature will often hover in mid-air, and it is not too difficult to snatch it away from the imaginary thread, by means of which it is suspended, by a lightning-swift movement of your hand. You will then notice that it buzzes almost indignantly—just as would a bumblebee. You are not taking the slightest risk, for like all other Syrphid flies this animal is perfectly harmless.

A second group of Syrphid flies is formed by those species whose larvae prey on aphids or plant lice. In most cases, the mature insects of these "aphid tigers" are wasplike, with yellow-and-black or white-and-black color patterns. Some are known as "crescent flies" because, on the back of their abdomens, they show a number of white, curved markings, standing out in a most lovely way against the velvety-black background of the body. They are fairly common, and I have found them in the Dutch lowlands as well as on the Alpine meadows of Mount Rainier in the state of Washington.

Probably the most abundant "aphid fly" in western Europe, however, is the currant Syrphid, *Syrphus ribesii*, which imitates a yellowjacket. Its American counterpart is *S. americanus*. It is really a great show to follow a mature female of one of these species when it is trying to deposit its eggs in the midst of a herd of aphids, such as you will sometimes find on roses, elderberries, sweet peas, beans, and other plants in your garden. Aphids are often used as "cattle" by ants, who like to sip up the sweet juice that emerges from their abdomen. When the *Syrphus* female approaches, the ants stop milking their aphid cows and begin to run around in great excitement, all the while threatening the fly. However, the performance always ends with a victorious Syrphid mother depositing her eggs on the leaf. The flat, wrinkled, green larvae that emerge from these eggs are very active, in spite of their blindness. While they hang on to the leaf with their bodies, their pointed front ends swing back and forth in search of victims, somewhat like the trunks of elephants. And indeed, every now and then you will see that a Syrphid larva manages to grab hold of a hapless aphid to suck out its juices.

In this book, however, we are really primarily concerned with the adult flies and their role as pollinators. Let me therefore mention that, in addition to the flowers of the Umbelliferae with their exposed nectar, there are still others, which are Syrphid favorites; for instance, the flowers of speedwell (*Veronica*). Most often we find that these are blue, of a hue that is particularly attractive to insects. In *Veronica Chamaedrys*—which is one of the commonest speedwells of all, found in both Europe and America as a weed in lawns—there is also a lovely honey-guide, a white spot in the center of the flower with dark lines on the petals leading toward it. A goodly quantity of nectar is found at the base of the pistil, and there is thus no doubt about the fact that this is an insect-flower. In the first two or three weeks in May, when this particular speedwell is in bloom, I love to watch the plants early in the morning. Each day, one or two, or even three, flowers in each inflorescence will open, in a process so rapid that it does not tax your patience at all. You can easily follow how the dark-blue petals unfold and how the two stamens spread apart wide, so that there will be one on each side of the flower, with the style in between. Since this *Veronica* is a humble little plant, the flowers are always close to the ground. Syrphid flies, therefore, have to approach from above, and in their efforts to find something to hang on to, they usually grab hold of the two stamens which will bend upward and inward. The result will be that the fly's belly will be powdered with pollen (Fig. 63). The next speedwell flower visited by the insect will, of course, receive this pollen on its stigma.

Another interesting point about this particular *Veronica* is that, toward the evening of the first day, the flowers begin to turn from dark blue to purplish. Since they do not fall off until the afternoon of the second day, one can observe in each inflorescence dark-blue and purple flowers at the same time. From what we have learned in other cases, we can safely say that "smart" insects, such as bees, quickly learn to distinguish between the two types of flowers. They will avoid the older ones which are poor in nectar and will thus save

a lot of time. We do not know whether this is also true of flies—
what a nice problem for an amateur to tackle—but since *Veronica*
is also busily visited by small wild bees, the color change would be
meaningful anyway. Now that I am on this subject of color again,
it is a striking fact that on the islands of New Zealand many types
of plants, such as gentians, which in other parts of the world have
blue flowers, appear with white blooms. It could very well be that
there is a connection between this phenomenon and the scarcity of
pollinating flies and bees on the islands. Did we not mention the

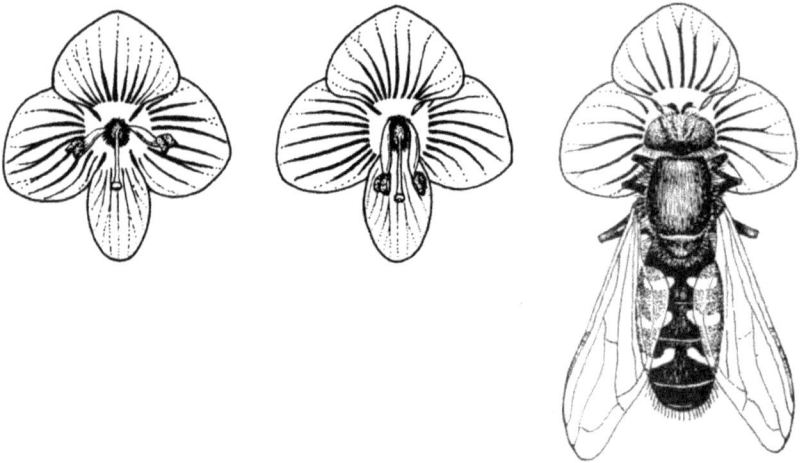

Figure 63. Flowers of a speedwell (*Veronica Tournefortii*) with crescent fly, *Syrphus lunulatus*.

story of the bumblebees and the red clover in Chapter 1? The
white gentians and other white flowers in New Zealand are probably pollinated by moths flying at dusk, when color is not important
but whiteness counts heavily.

From the point of view of the aphids, the flower flies which we
have discussed above are just plain parasites. The same can be said
for the beefly *Bombylius*, that "classical" superpollinator which we
have already mentioned in Chapter 3 in connection with Knoll's
experiments on color vision. It is also true for *Bombylius'* relatives.
Their larvae feed on the eggs and young stages of wild bees and
wasps, tiger beetles, grasshoppers, and some other insects. One of
the most fascinating of all is *Heterostylum robustum*, a parasite of
the alkali bee, *Nomia*, which we discussed in Chapter 11. The
females of *Heterostylum* act as real dive bombers and actually shoot

their eggs into the nest entrances of their victims. All you have to
do to collect quite a few eggs of the parasite is to stick a few empty
vials into the ground, open end up and flush with the soil surface.
*Heterostylum* will mistake these for *Nomia* nests and fill them up
for you.

Bombylids often are quite fuzzy, so that "bumblebee flies"
would be a more proper name for them than just "beeflies." We
have already seen that there is a fuzzy *Eristalis* too. However, the
most remarkable bumblebee imitators in the group of flower flies
are found in the genus *Volucella*. These not only look and act like
bumblebees but also deposit their eggs in bumblebee nests, and some
people believe their young are raised by their hosts; others think
the young act as scavengers or prey upon their hosts' larvae. The
color pattern of the hosts and the volucellas is said to be roughly
the same, so that a bumblebee species with, for instance, a red tip
of the abdomen will raise a *Volucella* species whose abdomen is,
likewise, red tipped; yellow-and-white bumblebees will raise yellow-
and-white volucellas, and so forth. A great deal of work, however,
remains to be done to establish this beyond doubt. Here we will
mention only one American *Volucella* that is a dead ringer for a
black-and-yellow bumblebee: *Volucella bombylans*, subspecies
*evecta*.

Now that we have encountered so many Syrphid flies that imitate bees, wasps, and bumblebees, it will be clear that the group
has played a very important role in the study of mimicry. The
theory is that in the struggle for life, the Syrphid fly has a definite
advantage in resembling a "model" which is shunned by natural
enemies such as birds. Or, in other words, imitating a wasp, a bee,
or a bumblebee has survival value. But we have to be careful; the
fact that, to us, a Syrphid fly and a wasp look the same does not
mean a thing, and the only way to arrive at valuable conclusions
is to do experiments with birds. Fortunately, a very good beginning
has been made, and at the same time some other questions have
been answered, such as the one concerning the possibility that there
is an *inborn* fear on the part of birds of yellow-and-black stripe
patterns. G. Mostler, who published his results in 1935, found that
young songbirds do not show such an innate fear at all. They consider Syrphid flies a delicacy, and this is true even for those forms
which closely resemble bees, wasps, and bumblebees. Wasps (yel-

lowjackets) were indeed unacceptable to Mostler's birds, which was partly due to the sting, partly to the taste of the internal organs of the abdomen. As soon as a young songbird had had a few sad experiences with wasps, it would leave them alone, and not only them but from then on also their imitators, the yellow-and-black, tasty Syrphids! It turned out that the memory of birds played a very important role. One young redstart refused a wasp eight months after it had last met one. The wasplikeness, then, although not immediately effective against young birds, begins to pay off for the Syrphid fly, as a species, after the bird has lived its normal life for some time and has encountered a few wasps. One way of putting this very neatly, although perhaps in an unconventional manner, is to say that the imitating species—in our case the Syrphid fly—must pay a certain tax in victims to educate every new generation of its predators, the birds.

W. Windecker, in 1939, confirmed these ideas very nicely by doing experiments with the well-known "zebra" caterpillars of the cinnabar moth, *Euchelia* (or *Hypocrita*) *jacobaea*. In western Europe, these insects occur in areas where ragwort, especially *Senecio jacobaea,* is abundant. I remember how we used to find big clusters of zebra caterpillars in the dunes near The Hague, where they would often do severe damage to their host plants. Again, young birds do not show any inborn aversion toward the larvae, with their alternation of black and yellow rings. They simply try to eat them, but reject them at once with such violent signs of disgust that it makes one laugh. What makes the caterpillars so unappetizing is their hairs. There are a fair number of these, but they are not very conspicuous. It is therefore not too amazing that the birds do not remember the hairs, but carry away with them the mental image of the color pattern, which from then on is associated with unpleasantness. They will refuse any insect with similar black and yellow stripes.

The next logical question in this type of experiment is: How close must the resemblance between model and imitator be to make it effective? H. Mühlmann has approached this problem in a very elegant way by painting meal worms, which are plain, yellowish-white animals, in various patterns of red bands and making them distasteful with tartar emetic. Birds soon learned to refuse such models, so that these now were safe from attack. Meal worms that had not been treated with tartar emetic were now given the same color pattern, or one which resembled it more or less closely, and the reaction of the birds toward these "imitators" was studied. As one would expect, a pattern identical with that of the model gave the best protection, but even a quite superficial resemblance was sufficient to be somewhat effective. The moral of this story, then, is that even a remote resemblance has some survival value.

There is no doubt that Sprengel, who obviously ascribed a certain consciousness to his insects, would have considered the imitation of wasps by Syrphid flies a neat trick. It is too bad he was not aware of it, for this might well have softened his judgment of "the dumb flies." My recording of the facts here can thus be construed as a belated act of justice. However, it is well to keep in mind that neither Sprengel's beloved bees nor the flies act in an intelligent way; the animals are not really aware of what they are doing. It is a good thing that modern investigators do not discuss them in terms of "dumbness" and "smartness" any more. This does not make the animals the least little bit less interesting, either.

# 15

# *The Ways of the Wasps*

When they discuss wasps, most people think only of yellow-jackets and hornets. This is quite natural, for these are wasps that form colonies and often occur in tremendous numbers. Moreover, since they have a sweet tooth, they frequently make their presence felt, bothering people in their quest for jam, honey, and soft drinks. Their life history is very much like that of the bees and the bumble-bees. This means that a young, fertilized queen begins to build a nest early in the spring, usually in March. Very soon afterward, the worker wasps that hatch from her first eggs begin to help her. They build a paper nest which they have to keep enlarging all the time, for from now on the colony grows very fast. In August, a wasp nest may count hundreds or even thousands of individuals, so that it really begins to resemble a beehive. However, when the nights begin to lengthen and the temperature drops, the colony gradually dies out, with the exception of some young, newly ferti-lized queens. These stay in a snug hiding place during the winter, to emerge on one of the first fine days of the next spring.

Strictly speaking, there is only one real "hornet" in America, *Vespa crabro*, introduced from Europe to New York around 1850. It has not spread very far as yet, and I cannot say that I feel sorry for that, for this is really a vicious wasp which does not hesitate to attack people and large animals. In Schiller's *Wilhelm Tell* you can read how the horse of a knight carrying an important message was killed by a swarm of these hornets. In general, it is wise to give a wide berth to the large state-forming wasps. This point was driven home to me very cruelly when the mother of one of my

156

playmates in the East Indies, on a trip to a mountain crater, was attacked and killed by a swarm of them, in country where it was impossible to find any shelter from the crazed animals.

In *V. crabro*, some individuals are more than an inch long. The color is reddish brown with dull yellow markings. These wasps use living bark and wood for building their brittle brown paper nests and can, therefore, do a lot of damage to trees and shrubs. Most often, the nests are to be found in hollow trees. They are quite big, sometimes more than three feet high, with layer upon layer of paper combs, each with hundreds of cells. When I was a student, we liked to watch the entrance hole of such a nest through field glasses, from a safe distance, hoping to see the "trumpeters" or "buglers" at work. Tall yarns have been spun about these mysterious creatures, but they do exist; they act as ventilators, producing a whirring sound. We also liked to see our hornets—and yellowjackets, for that matter—at work on "bleeding birches." These are trees attacked by the caterpillar of the goat moth, a huge animal (as caterpillars go) that tunnels for 3 years in the wood before it pupates. From the wounds in the tree trunk, a sweet sap oozes out, which soon begins to ferment, attracting dozens of butterflies and other insects.

The wasps certainly do not visit these trees for the birch wine alone. Every now and then, one of them will pounce upon an unsuspecting fly or a butterfly, bite off the wings of the helpless victim, and take it home as food for the wasp larvae. Yes, this is one of the great differences between bees and wasps. Meat is the only food these larvae get, and no worker wasp will ever take pollen or nectar back to her colony. In this connection, it is quite revealing to catch a wasp in a glass tube and to examine with a hand lens the few hairs she has on her body. Whereas bees possess numerous hairs that are branched, or even twisted in a corkscrew fashion, so that they are marvelous instruments for the collection of pollen, the hairs in this case are simple and straight. This is what one might expect, because as far as wasps are concerned, pollen is just a nuisance. They are only interested in flowers for the nectar, which they consume on the spot.

It can be understood at once that, because of this difference, wasps just do not compare with bees as pollinators. Every day, a hive bee will collect at least twenty-five times as much nectar as she needs for her own consumption. She will, therefore, visit hundreds upon hundreds of flowers, whereas for a wasp a few dozen must be considered a good score. However, hornets and yellowjackets can become quite numerous—there are very pronounced "wasp years"!—and they will make up by sheer weight of numbers for what they lack in efficiency.

There are some flowers for which wasps show a definite preference, and which in their turn seem to rely somewhat on wasps for their pollination. I remember, for instance, that in Holland there are three wild orchids (*Epipactis* species) that are always referred to as "wasp orchids." Some scientists maintain that the brownish-purple color which we notice in their flowers has a special attractiveness to wasps. Although this has never been proved conclusively, I do know that one can frequently find a dozen of these animals on a single orchid plant. Each flower offers its nectar in a neatly shaped little cup, a depression in the upper part of the "lip." The wasps plunge their heads into the glittering, sweet fluid without hesitation, and lap up the nectar with all their might, always moving their mouth parts in an *upward* direction. This favors the chances for pollination in an excellent way, for directly above the little cup is placed the squarish stigma, the upper part of the pistil, and immediately above that organ, fused with the back part of the very short style, we find the lone stamen which actually is just an anther. In each compartment of the anther, all the pollen grains stick together, forming a "pollinium" which in front tapers out into a little stem tipped by a sticky, shiny, tiny blob. In her greedy lapping movements, the wasp just cannot fail to touch that little blob; it will stick to her head, and when she pulls back her body the two pollinia will follow. For a few brief moments, they stick out straight from the wasp's forehead, but before long they start bending forward a little, so that when the wasp visits the next flower they are in an ideal position to be pushed against the stigma. Sometimes, however, the wasp just continues to drink nectar from one and the same cup for a long time. In this case, the bending-through of the pollinia will result in self-pollination. Somehow this does not seem to affect the quality of the seeds, for when fall arrives we usually find that all the flowers have formed nice fat seed pods with innumerable tiny seeds in them.

One of the European *Epipactis* orchids, introduced accidentally into this country, has been fairly successful in some of our eastern states, in Pennsylvania, for example, where one can find specimens growing as weeds along roads and railroad tracks. However, there are some original American *Epipactis* species, too, for instance the tall stream orchid, or "chatterbox," *E. giganteus*, which is very fond of water and is often found growing in brooks and rivers in California. It might certainly be worth our while to find out if wasps are also active there.

Of the wasps that do not form states, we have already mentioned the bee wolf, *Philanthus triangulum*, in our discussion of the honey-guides. Although this wasp does visit flowers, the females probably get most of their nectar by sucking it out of their paralyzed victims, the honeybees. They do need a lot of energy-rich food because they have to transport the bees over large distances—not an easy job when we consider that robber and victim are almost equal in size.

Another fascinating wasp which one sometimes finds on flowers such as fireweed and dodder is the caterpillar-hunter, *Ammophila*. One could not ask for a thinner wasp waist than the one seen in this animal. As a matter of fact, the whole creature is extremely slender, with the exception of the tip of the abdomen. Red and black are favorite colors in *Ammophila* wasps. My friend, Professor G. P. Baerends, has made a beautiful, intensive study of the western European caterpillar-hunter, *A. campestris*. For days or even weeks on end, he and his wife followed the adventures of individual, marked, female ammophilas around their nesting-sites. In this way they discovered the almost incredible fact that the wasps can take care of several nests with brood at about the same time, but with considerable overlap, providing the larvae with the right sort of food at the right moment. It is just as if the wasp has, in her little brain, some sort of foolproof bookkeeping system. Of course, this story has little to do with pollination, but it is so remarkable that I just cannot resist the temptation to tell you about it. Here, then, is the *Ammophila* story as Baerends recounts it:

The wasps emerge from their cocoons in June. After having mated, the females start digging their nests. They prefer sandy soil and flat surfaces such as one finds on paths, and dig down to a depth of a little over an inch. A nest simply consists of a vertical tunnel with one little chamber or cell at its bottom. The cell is provided with paralyzed caterpillars. An *Ammophila* egg is attached to the first victim, and the caterpillar serves as food for the larva which hatches from it.

In the behavior of the female *Ammophila*, three phases can be distinguished very clearly. In the first of these, the wasp digs a nest, stocks it with one caterpillar, and deposits her egg. The beginning of the second phase is marked by a visit of the mother wasp, during which no fresh caterpillars are brought in. This we shall call the empty-handed visit. If, at this moment, the cell turns out to contain a larva—which means that the egg has hatched—the mother wasp will follow up her visit by bringing in one to three caterpillars. The third phase starts with another empty-handed visit, followed by further stocking of the nest with three to seven caterpillars. After this, the nest is closed with minute care and left to its fate. When the female *Ammophila* has completed a certain phase in one nest, she will go to another to complete the phase that is in order there. After having completed her job in this second nest, she either goes back to the first or starts on a third.

She will always do the job demanded by the situation in the particular nest she happens to be working on. Thus, a cell containing an egg is not stocked with more caterpillars; a young larva, as we have seen, gets the one to three caterpillars it is entitled to, and an older larva just the right quantity of three to seven caterpillars. How does the female wasp do it? In order to find this out, Baerends replaced the natural nests by nests made out of plaster of Paris. The females continued to take care of the brood in these as if nothing had happened. In the artificial nests, the contents of the chamber could be changed at will. It turned out that the empty-handed visit which marks the beginning of the second and third phase is of crucial importance. Whether or not *Ammophila* will bring in additional food is determined by the contents of the nest chamber at the time of that visit! She will only bring in something if she finds a larva present, not an egg; the *quantity* of caterpillars to be brought in is determined by both the size of the larva and the quantity of caterpillars already present. Because the female *Ammophila*, when she has completed a certain phase, will always pay an empty-handed visit to a nest which still needs care, and also because she does not start a new nest as long as any of the existing nests need food immediately, each of her broods gets exactly what is necessary. Depending on the situation she finds during her empty-handed visit to a given nest, she may start on the second or third phase in that nest. However, if she is not stimulated to bring in additional food, for example, because the egg has not yet hatched, she starts out on the first phase of a new nest. With the aid of this wonderful regulation mechanism, each female takes care of a whole succession of nests, until the beginning of September.

Wasps are also involved in the strangest case I have to record in this book—a case so incredible that I would not accept it myself if it had not been reported by at least half a dozen reliable investigators on three continents. The opposite numbers of the wasps in this affair are orchids. From Darwin's little masterpiece on these plants, it is quite clear that one can expect almost anything from them, as far as pollination is concerned. Yet, even Darwin himself might, for once, have been left breathless had he known about the things we are going to discuss here.

There is in southern Europe, and also across the Mediterranean in Algeria, a queer little terrestrial orchid known as the "looking-glass *Ophrys*," *O. speculum*. Its flowers are nectarless and rather small, yet striking. In 1870, a British flower-lover remarked that "the brilliant polished surface of the disk of the lip, which shines like a blue-steel looking-glass, edged with gold, and that again set in a rich maroon velvety frame, presents a combination of colors quite unlike anything else known to me in the vegetable kingdom." Indeed! We expect metallic colors in the world of the insects, not in flowers. Their significance in *Ophrys* was not made clear until 1916. At that time, after more than 20 years of patient work in Algeria, A. Pouyanne published his conclusion that the flowers of *O. speculum* imitate the females of a wasp, *Scolia ciliata*. The violet-blue center of the lip gives, indeed, the same optical effect as the reflections from the halfway-crossed wings of a resting female. A thick fringe of long, red hairs which sets off the yellow margin of the lip imitates the hair fringe found on each segment of the insect's abdomen. The short feelers or antennae of the female wasp are beautifully reproduced by the upper petals of the orchid which are quite dark and threadlike. And so on; careful observation reveals several more traits shared by insect and flower (Fig. 64). Admittedly, the resemblance is not perfect, but it is sufficiently perfect to attract *Scolia* males, who will treat the *speculum* flowers as if they were their females. This behavior of the males, known as pseudocopulation, leads to the withdrawal of the pollinia from the flower visited and to cross-pollination later on. By cutting off the

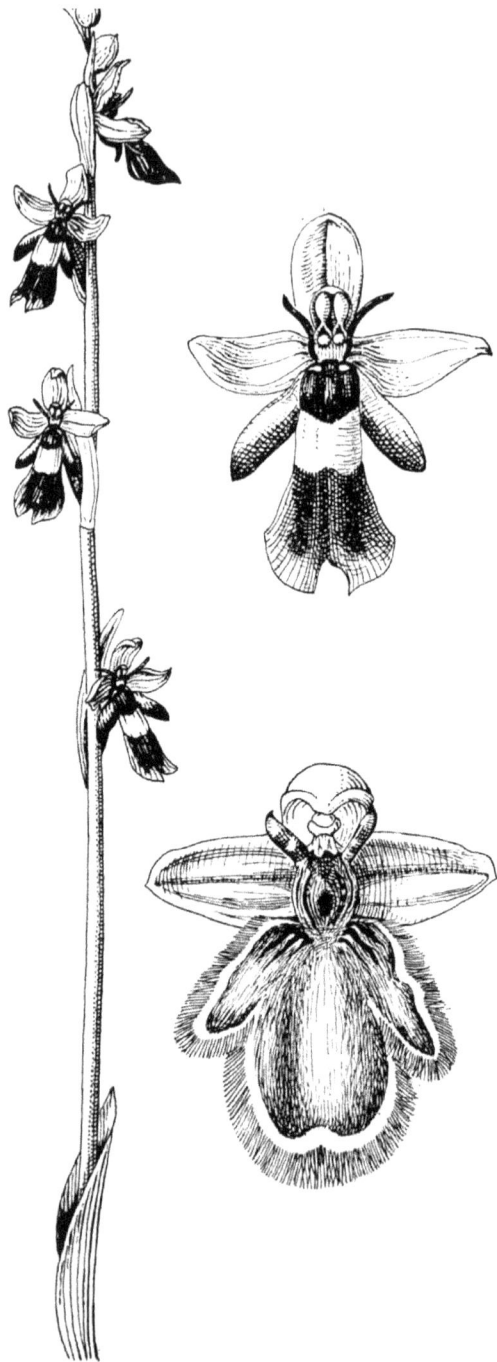

Figure 64. Pseudocopulation in the terrestrial orchid, *Ophrys*. The flowers of some species imitate female wasps or bees and are treated as such by the corresponding males, a situation which leads to cross-pollination. Left: flower spike of *Ophrys muscifera* Huds. from western Europe and Scandinavia. Upper right: individual flower of *Ophrys muscifera*, enlarged. Lower right: flower of the looking-glass Ophrys, *Ophrys speculum*, from southern Europe and Algeria.

labellum and concealing the speculum flower in newspapers, Pouyanne could show that both sight and smell are important in attracting the *Scolia* males.

It can be shown in still another and very elegant way that the *Ophrys* flowers are good imitators of wasps. If we cut them off and attach them to such insect-beloved flowers as peonies and bryonias, we will find that bees and bumblebees begin to shun these. So, the *Ophrys* flowers, or even their detached lips, act as bee repellents. The reason is that the bees and bumblebees have a natural tendency to avoid flowers already occupied by visiting big insects. It is clear that they mistake the attached flowers for such insects. Even *Ophrys* flowers left in their natural position, on their spikes, are rarely visited by bees and bumblebees, although in a given area these insects may be present in tremendous numbers. Probably the bees mistake the *Ophrys* flowers for insects resting on a leafy green stem or for small green flowers with a large insect already sitting on them.

For *O. speculum* as well as for other *Ophrys* species, Pouyanne's results have been confirmed in a beautiful way, first by M. Godfery in southern France and later by K. Faegri, B. Kullenberg, and T. Wolff in Scandinavia. The most extensively studied *Ophrys* species nowadays may well be *O. muscifera* from west-central Europe and Scandinavia, which is pollinated by the digging wasp, *Gorytes mystaceus*. The *O. lutea* and *O. fusca* from southern Europe are pollinated by male andrenas (wild bees). In this case, the insects enter the flower with the abdomen first.

It is most gratifying that a completely independent investigator in Australia, Mrs. Edith Coleman, has come up with a very similar story, although a different type of wasp and different orchids are involved. *Lissopimpla semipunctata*, a slender ichneumonid wasp, is responsible for the pollination by means of pseudocopulation in four species of *Cryptostylis* (*C. leptochila*, *C. subulata*, *C. erecta*, and *C. ovata*). In these orchids, the slender flowers do an excellent job of imitating the antennae and the ovipositor of the females; even the white spots on the female abdomen are imitated by means of a double row of glistening glands on the lip. Figure 65 shows a *Lissopimpla* male with *Cryptostylis* pollinia attached to the tip of his abdomen. It is quite clear that the insect was not interested in food, because in that case the pollinia would have been attached to a different part of his body.

Gall wasps lead a life entirely different from that of the free-roaming beewolves or the swashbuckling hornets. Their development takes place completely in the interior of plants, sometimes in the leaves, sometimes in other parts such as the roots or the flowers. The female gall wasp deposits her eggs in a certain spot,

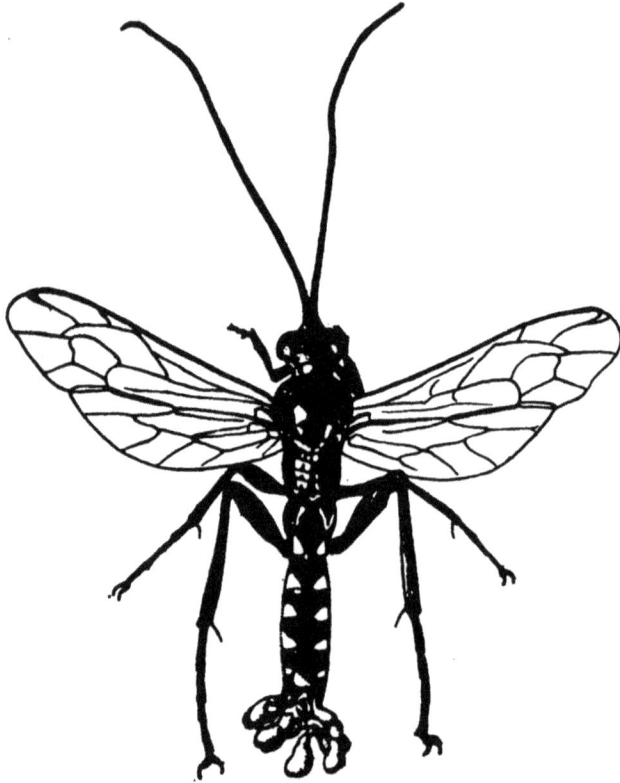

Figure 65. A male specimen of the ichneumonid wasp, *Lissopimpla semipunctata*, caught after having visited the flowers of an Australian orchid, *Cryptostylis*. Obviously, the wasp mistook the orchid flowers for females of his own species; the pollinia are attached to the tip of his abdomen.

and the plant tissues around it respond by forming a "gall," which grows bigger with the developing larvae; these young animals stay right where they are. Galls come in all shapes and colors. Well-known ones are those on oak leaves which resemble small red-cheeked apples, and the galls on roses which look somewhat like delicate corals.

There is an important group of plants where gall wasps are the sole pollinators. This is the group of the figs (*Ficus*), which has about 700 different species, most of them tropical. I think that many of us have seen photographs of the notorious "strangler figs," which take away so much light from the host trees on which they developed that they eventually kill them. The sacred banyan tree of India, the waringin of Indonesia, and the rubber plants which we grow in our homes as an ornamental are all figs. But the most important *Ficus* is the Mediterranean fig tree, *F. carica*. The ancient Greeks were already familiar with the active role played in this plant by a small gall wasp, *Blastophaga psenes*, and applied their knowledge in practice.

In order to explain the situation in figs adequately, I must first tell you that what we call a "fig" actually represents a complete inflorescence. The ancestors of the fig trees probably had inflorescences very much like those of the present-day nettles, which are related. Each inflorescence had a great many flowers. Imagine now, that all the branches of such an inflorescence began to spread out in a horizontal plane, and that flowers were formed on their upper sides only. Assume, furthermore, that the branches became so fleshy that they fused sideways with their neighbors; finally, imagine that the thick flower-bearing plate thus formed became a shallow, fleshy cup and then a deep beaker with the flowers placed on the inside. It is clear that—after all this—we have obtained a fig. That such a process must actually have taken place is shown by the fact that nowadays we still find certain plants, dorstenias, with inflorescences about halfway between those of the nettles and those of the figs. The hard, tiny grains which we notice when we chew on a ripe fig are the stony kernels or seeds of the individual fruits scattered in the fleshy part of the beaker.

In certain parts of Italy, wild fig trees, known under the Latin name *F. carica erinosyce,* can still be found today. They play host to little *Blastophaga* wasps that take care of their pollination, and it can be honestly stated that there exists a beautiful symbiosis between the two partners. The timing between the life history of the gall wasps and the development of the fig tree's fruits is perfect, and if one partner should for some reason die out, the other would inevitably follow. These wild figs are monoecious, which means that both male (or staminate) and female (or pistillate) flowers are formed on one and the same plant. Moreover, flowers of a third type are sometimes found, the so-called "gall flowers" which are modified female flowers with a very short style.

In the spring, the cycle starts when a wild fig tree starts forming so-called "profichi," that is, inedible, medium-sized figs which contain, close to their entrance, a number of male flowers and farther down, on the sides and bottom of the beaker, numerous female flowers with short styles. Female gall wasps penetrate these profichi to deposit their eggs in the latter type of flowers. Soon afterward, the insects die. The flowers containing their eggs, however, change into galls, each with a larva inside.

In June and July, when the profichi reach maturity, the wingless male wasps and the winged females hatch. Each male frees himself from his prison by chewing a hole in the hard wall of his gall, finds himself a female in the same fig prison, and then chews a hole in the wall of her prison, too. After mating has taken place, the female enlarges this hole, crawls out of her gall, and leaves the fig where she was born. In doing this, however, she has to cross the region of male flowers near the fig's entrance, which have just opened. Powdered with their pollen, she will now make her way to the young edible figs ("fichi") which have just formed on the same tree or on another one close by. The fichi contain, in large numbers, genuine female flowers with long styles, able to resist successfully the repeated attempts which the female wasp makes to deposit eggs in them. In these attempts, the flowers will be pollinated, and it is not unusual to find that one female can take care of all the flowers on a fig.

The beaker's entrance is now gradually closed because the scales which are present here become much bigger, the seeds mature, and the whole fig turns into a fleshy and tasty sweet fruit. Long before this, however, the unsuccessful female has left the immature fig to continue her efforts in another one of about the same developmental stage—but again in vain. She cannot satisfy her urge to deposit eggs in fig flowers until a third type of fig has begun to develop on the trees, the so-called "mamme" or mother figs. They are formed early in the fall and only on the upper branches. In their interior, these small, inedible figs contain only gall flowers, excellent incubators for the eggs and larvae of the female wasps. The generation of young wasps produced in the mamme figs hibernates in the larval stage, so that males and females do not emerge and mate until the next spring. After that, the profichi–fichi–mamme cycle is repeated.

The seeds formed in the fichi of the wild, Mediterranean fig trees are honest-to-goodness, viable seeds. After the fichi have been eaten and digested by birds or other animals, they can be deposited with the excrements in all kinds of places, such as crevices in rocks or walls, where they will give rise to new *erinosyce* figs.

A question arises from what we have just presented. How did we get our present-day fig trees, which are almost completely female and which produce edible fruits several times each year? We must assume that thousands of years ago, probably in Asia Minor, the ancient Greeks began to grow fig trees from cuttings taken from the wild *erinosyce* trees of those days. They were fortunate enough to achieve a "separation" and to end up with two different types of trees, one producing essentially only figs of the "profichi" type and the other yielding good edible "fichi" or figs. The first type of tree is known as "caprifig" (*caprificus*). In spite of the fact that its products are inedible, the *caprificus* is, of course, invaluable as a producer of pollen and as a nursery for the gall wasps. For all practical purposes, it can be considered a male tree. This was recognized by the ancient Greeks and Romans, who planted the caprifigs in the vicinity of the female trees to facilitate the job of the insects and to insure a high yield of edible figs. For the same purpose, cut-off caprifig branches bearing "gall figs" were sometimes attached to female trees. Even grafting of caprifig branches onto female trees was occasionally practiced. The systematic use of gall figs by fig-growers was known under the name of "caprification." Unfortunately, the principle was lost sight of or neglected when Smyrna figs—females—were introduced into California around 1880. The trees remained sterile for years, and did not begin to yield a good crop until after the introduction of the corresponding caprifigs and gall wasps! I have to admit that, nowadays, varieties of high-quality consumption figs have been developed which do not require the collaboration of insects. Needless to say, these figs are seedless, and have to be reproduced by means of cuttings. Caprification is still needed for those varieties of fig trees from which dried figs of good keeping quality are obtained.

# 16

# Flashing Beauties and Dashing Flyers

When I tell my complaining friends that butterflies are not at all rare in our area, they give me that saucer-eyed look of disbelief. So I have to explain to them, gently, that they do not have the right approach. What do you do if you want to get to know your neighbors? You throw a party, of course! For the butterflies, it is just the same: to attract them, you have to offer them the flowers which they really like.

Probably the very best way to cater to butterflies is to plant *Buddleia* or butterfly-bush in your garden. Some people call this plant Japanese lilac, and when it is in flower you can easily see why. True, the real lilac is more pleasing to the eye; *Buddleia* inflorescences are often longish and thin and somewhat crooked, and the whole plant gives an impression of untidiness. But then, what an abundance of flowers, at a time (late in the summer) when you can really use some color in your garden! And what a delicious fragrance! The purple of the *Buddleia* flowers is made even lovelier by the scores of butterflies that visit them: whites and blues, coppers, skippers and red admirals, painted ladies and mourning-cloaks, tortoise-shells, swallowtails—a living butterfly collection. On sunny days early in September, it is not a rare thing at all to find more than a hundred individuals on one single *Buddleia* bush. They have probably spent the cool night in some brush heap or against an ivy-clad wall where they were protected from the wind. Around 10 A.M., when the sun's rays begin to be effective, they become active again and start drifting toward the rich table set for them. Most probably, it is the *Buddleia* fragrance that plays the main role in attracting butterflies.

Standing motionless near a *Buddleia* bush, it is easy to observe the insects while they are feeding. Perched on the inflorescences, the butterfly unrolls its long tongue or proboscis (Fig. 66), which in the rest position is kept coiled up like a watch spring, and sticks it into a flower. Sucking continues for several seconds, and then another flower is tackled. Every now and then, a butterfly will

Figure 66. A sulphur butterfly, *Colias*, sipping nectar from a marjoram flower. Notice the uncoiled proboscis.

take a few steps, and now you can see that there are some which have six legs, which is what the books claim for all insects, and some with only four. The swallowtails and the whites are in the first group, and the close relatives of mourning-cloak and tortoise-shell, the nymphalids, are in the second. In Chapter 3 we have seen that these two groups also differ in their color vision: whites and swallowtails see red as a color, nymphalids don't.

Talking about legs, I would like to draw your attention to the red admiral, a butterfly that is found the world around in the Northern hemisphere and visits *Buddleias* often. It is hard to confuse it with any other butterfly, at least so long as it keeps its wings open, which it does most of the time. The general color in this pose is a very dark, velvety brown; the hind wings have a red border, the front wings an oblique red band about half way between tip and base. Moreover, there are some irregular white dots near the wing tips. As I mentioned in Chapter 9, this butterfly can taste with its feet, and can distinguish between pure water and sugar solutions which are so extremely dilute that we humans do not notice any sweet taste at all. In this respect, the red admiral is very much like the bluebottle fly, *Calliphora*, which also "has its tongue in its toes." It is not surprising that red admirals are so often found on the sweet sap that oozes out of bleeding birches and oak trees; sometimes you find a dozen of them on a single tree.

I am not sure what the situation in other butterflies is, but I would think that there are several others which can taste in the same manner. The first one to suspect is, of course, the red admiral's closest relative, the painted lady or thistle butterfly. At first sight you might doubt the kinship, because both the general color (a warm orange-brown) and the pattern (numerous small black spots) are so different from those of the red admiral. However, the shape of the wings, at least, and the arrangement of white dots at their tips, is very much the same. The painted lady is rumored to be the most widely distributed of all butterflies. This I am quite willing to believe, for I remember having found them often enough when I was a little boy in the East Indies. It is possible that the wide distribution has something to do with the food plants of the caterpillars, which in most places are quite common: nettles, thistles, hollyhocks, sunflowers, burdock, and some others. The caterpillar is greenish gray, with dark spots and numerous light, branched spines. In contrast to the caterpillars of mourning-cloaks and tortoise-shells, which you find together in large groups, those of the painted lady are rugged individualists. Each lives alone, for example, in a nettle leaf rolled together very neatly. This may be another manifestation of the relationship with the red admiral, for the latter's caterpillar does exactly the same thing. Painted ladies often occur in countless numbers. In the summer of 1958, their

caterpillars could be found all over the place in the Pacific Northwest. They migrate to other areas in swarms that may contain millions of individuals.

Another migrant is the monarch or milkweed butterfly, which is
so well known to the American public that it might well be called
the national butterfly. In the fall, large numbers of monarchs from
the cold North will assemble in certain spots in more hospitable
regions, where they spend the winter. Year after year, they select
the same locality for this, and even the same trees. Probably the
best-known of these hibernation sites is in the town of Pacific
Grove on Monterey Peninsula in California. The arrival of the
monarchs in the fall is an event that is celebrated every year. The
butterflies are protected by law, and form a tourist attraction of
the first order. They sit on the trees in large clumps, or rather
garlands, practically touching one another, so that the trees seem
to be decked out in an exotic foliage of the most gorgeous colors.
These butterflies are not in a complete state of topor; every now
and then, one will fall down from a clump to sail around on its
wings in a lazy, majestic manner. There must be several such
hibernation sites in the western and southeastern United States, but
none of them, probably, is quite so large as the one in Pacific Grove.
In the spring, the monarchs begin their long northward trek. Wher-
ever they find milkweed plants, they deposit their eggs, so that a
new generation will develop. A goodly number of monarchs will
end up in Canada.

Although experiments in which migrating monarchs were marked
have not been too successful, there is good reason to believe that
the butterflies that come south in the fall are young individuals
which have never before seen the hibernation sites. What enables
them to find these is still one of those elusive mysteries of Nature.

There are many other things besides its migration which make
the monarch one of the most interesting of butterflies. For instance, it has often been found in Europe, and this has led to quite
a controversy. Do these butterflies cross the Atlantic on their own
power, or are they transported passively on ships? Most scientists,
nowadays, are inclined to accept the latter explanation. The largest number of European monarchs has been reported from England,
which maintains a steady boat traffic with America, whereas only
a very small number has been reported from Spain and Portugal.

If the animals could fly across the Atlantic, the numbers should at least have been equal. I should add that they soon die out in Europe because their food plant, milkweed, is lacking or at least very uncommon there.

Another remarkable fact about the monarch is that it has an imitator, a butterfly which very appropriately has received the name of "viceroy." This is another example of mimicry. Just as in the case of the bees and the "blind bees" which we discussed in an earlier chapter, the imitator or mimic is quite good-tasting, whereas the model (the monarch) is almost completely unacceptable to birds. Although they certainly had good reason on their side, scientists have for a long time only been theorizing about the monarch-viceroy situation. Fortunately, ironclad proof has now also been provided by Jane Van Zandt Brower in a neat series of experiments with Florida jays.

Before I leave the monarchs, I would like to point out that it is the easiest thing in the world to raise the caterpillars of these butterflies, with their zebra-like cross-striation of black, yellow, and white lines. It is fun to watch them while they are eating. When you or I damage a milkweed leaf with its veins full of latex, this milk juice will flow out immediately. The caterpillar, however, must have means to prevent this, for it is never embarrassed or engulfed by an overabundance of sap. The chrysalis of the monarch is a marvel of beauty, a green jewel with little nails of gold.

The fact that monarchs hibernate should destroy the belief which many people still have that the life span of butterflies is just one day. You can find out for yourself that monarchs are not an exception by marking a few other butterflies on your *Buddleia* with a dot of quick-drying paint. If you are lucky, some individuals will return day after day. Even when the flowering of the Japanese lilac is over, they may still be around, visiting dahlias, asters, and goldenrod. They will disappear only gradually, as the weather grows colder. However, a careful search would reveal that the mourning-cloaks, the tortoise-shells, and several of their relatives were not dead; they had found shelter in nooks and crannies of the same sort as those in which they used to spend their nights. This is simply their way to hibernate. Certainly, many of them will be killed by mice and titmice and some perhaps by the cold. Nevertheless, in February or March a few survivors will emerge to bask in the sunshine again and to feed on the first flowers of spring. A wonderful plant to attract these early birds is *Daphne Mezereum,* originally a wild shrub from central Europe but nowadays cultivated in many gardens in the temperate zone. The delicious fragrance of the purplish-pink *Daphne* flowers is very hard to describe and is very effective.

The wildflowers visited by the spring butterflies are roughly the same ones frequented by bees and bumblebees. The various coltsfoot species (*Tussilago, Petasites*) seem to be very special favorites, but even willow trees receive their share of butterfly visits. The females of the mourning-cloak also select these trees and birches on which to deposit their eggs. They die soon after having done this, but that still gives them a life span of almost a year.

Later in the spring, the butterfly army is reinforced by those species that have spent the winter in the chrysalis stage. In this group are most of the swallowtails and the whites. Among the latter, the so-called "orange tips" (*Anthocharis, Euchloe*) are just about the prettiest and most interesting. With their wings open, they are rather conspicuous little butterflies, especially the males. When resting or feeding, however, they fold their wings together and display only the underside, which is marbled in the most delicate green and whitish tinges, giving them a magnificent camouflage. The orange tip caterpillars are to be found mostly on plants of the mustard family, Cruciferae, and it builds up our sense of fair play when we notice that the adults in their turn are helpful in the pollination of the flowers of these.

Before we devote too much time to the animals, however, let us examine the characteristics of the typical butterfly flower. Some of the wild pinks are good examples. They are often red or pink, which cannot be just a coincidence; it must be remembered that swallowtails and whites, in contrast to bees, can see red as a color. The petals taper down into a very slender, pointed base while the long, tube-shaped calyx which holds the bases of all the petals together is, in its turn, surrounded by some extra "bracts" so firm and solid that they thoroughly discourage any bumblebee which might be tempted to burglarize these flowers in an effort to get at the nectar in the bottom part of the tube. Thus, only insects with a long proboscis can reach the nectar and these are, of course, mainly butterflies. The flowers of the pinks really need their help, for they are proterandrous: on the first day of flowering, five mature anthers are visible in the flower entrance; on the second day, these are replaced by five others; and it is not until the third day that the receptive stigmas finally take their turn.

Other butterfly plants are the purple loosestrifes (*Lythrum salicaria* and close relatives), found so abundantly in wet places. They are fairly tall plants, often occurring close together over rather large areas. From a distance, therefore, their flowers seem to form one huge carpet, which in the summer may set off certain lakes and streams in a most lovely way. The seeds which the loosestrife flowers produce when they undergo self-pollination are not too good, and the help which the butterflies provide is, therefore, most valuable. However, the situation is far from simple. The following explanation may be helpful.

Loosestrife flowers always possess twelve stamens, organized in two groups of six which differ in length; within each group, the length of the stamens is quite uniform. Inspection of a large enough number of flowers, however, shows that the stamens come in three sizes: very long, medium, and short. From this it follows that there must be three types of *Lythrum* flowers: (1) a type with six long- and six medium-length stamens, (2) another with six long and six short stamens, and (3) a type which has six medium-sized stamens and six short ones. There is only one pistil per flower, in *Lythrum*, but this organ, too, can be very long, medium-sized, or very short. Flowers with stamens of type 1 always have short pistils, those of type 2 have medium-sized ones, and those of the last type have long-styled pistils. When, finally, I explain that the color of the pollen and the filaments may differ, too, so that short stamens have purple filaments and yellow pollen, and long stamens have white filaments and green pollen, you may begin to think the *Lythrum* flowers present a sorry mess indeed. But once you have the key to the situation, the solution is simple. An abundance of good, viable seed is produced only when pollen from the long stamens goes to the high-placed stigmas of equally long pistils, or when pollen from medium-length stamens finds its way to medium-length pistils, or, finally, when short-stamen pollen goes to low-level stigmas. Essentially, the situation is not different from the one we have come across earlier in primroses. There, however, we had only two "good" combinations, and here three. Figure 67 may help to make things a little easier to understand. It is clear that a great deal of work has to be done by the insects here, and it is therefore a good thing that all kinds of butterflies (swallowtails, whites, blues, brimstone butterflies, fritillaries, and so on) are very fond of *Lythrum*.

A third group of flowers adapted to pollination by butterflies is formed by the violets, or perhaps it would be better to say by some violets, for they are not all the same. These flowers have a spur, formed by one of the petals. The nectar which accumulates

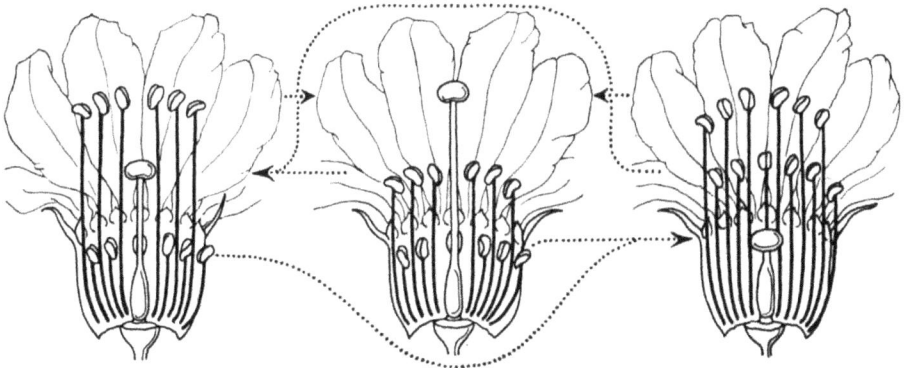

Figure 67. The three flower types of *Lythrum salicaria*, the purple loosestrife from western Europe. Notice the three possible levels of stamens and pistil. Ideal pollinations are indicated by arrows.

in this spur is secreted by two of the five stamens; each of these two stamens has an appendage that sticks out into the spur. It is hard to see the stamens unless one dissects the flower, for they are very short, with thick anthers whose tips are provided with appendages. Anthers and appendages together form a cone which surrounds the style very closely. The stigma, which often is handle-shaped, sticks out from this cone. Since the stems of the flowers in violets are curved, anther cone and stigma point *downward*. The anthers release the dry pollen on the inside, into the cone, where for the time being it stays. However, when a visiting butterfly sticks its proboscis into the spur to get at the nectar, it must inevitably touch the stigma. The anthers move apart a little bit, and the pollen rains down on the proboscis and into the spur. When, after this, the insect visits another violet flower, the pollen will be deposited on a receptive stigma. Even without insects, one can very easily find out how things are brought about in violets, by using a very thin piece of wire. Figure 68 may be of some help, too.

In addition to butterflies, bees, bumblebees, and beeflies have been observed as efficient pollinators on violet flowers of the type just described. *Viola calcarata*, however, a form which is found in the Alps, has a spur which may reach a length of a full inch. It is,

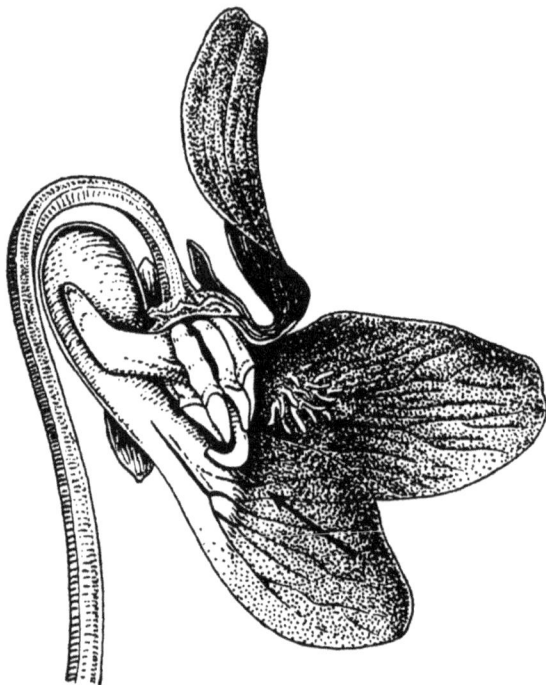

Figure 68. Section through the flower of a violet, *Viola odorata*. The nectar is secreted into a spur by two modified stamens. Insects such as butterflies, trying to get at the sweet fluid with their proboscis, cannot fail to touch the column formed by the pistil and the other stamens, which acts as a salt-shaker because the pollen grains are released on the *inside* of the column and come out as soon as the free end of the pistil is touched.

therefore, safe to say that this species is entirely dependent on butterflies for its cross-pollination. Often, these butterflies are fritillaries. There is some poetic justice in this, for the caterpillars of the same fritillary species that achieve pollination may live on the violet's leaves. These fritillaries are in the group of the meadow-browns, which is closely related to the nymphalids. Like these, the meadowbrowns have only four legs. The first of the three pairs of legs which butterflies should rightfully have, has become very small and hairy and is used only for sprucing-up purposes.

Although the meadowbrowns, as a rule, are more modest in color than the dazzling nymphalids, they are at least as fascinating. For instance, when you compare the males of some of your violet fritillaries with the females, you will notice that the males have a number of dark lines on the upper surface of their front wings. Under the microscope, it turns out that the minute scales which form these lines are different from the ordinary wing scales of butterflies and have the shape and structure of a paintbrush with a hollow stem. These are *scent scales*. At their base, they were originally connected with certain glands that secrete substances of a delicious fragrance. When they court the females, the males go through quite a ritual. Facing their partners, they perform a kind of dance which always ends up with a low bow. The front wings, which at this stage of the game are kept just a little bit apart, are pushed so far forward that at least one of the feelers of the female—the organ that can perceive smells—seems to touch their upper surface. We cannot escape the conclusion that the purpose of the whole ceremony is to influence the female very strongly with the flowery fragrance of the male.

Actually, fritillaries are not the best animals for studying the "perfume dance." In an earlier chapter, I have already presented the grayling, a European butterfly for which we have coined the name "bark with wings" because of its incredibly good camouflage. In the males of the grayling, the scent scales on each forewing are united in a single dark patch. The odor which they produce is somewhat like the fragrance of cocoa. By removing the scent scales and covering the scent patch with shellac—innocent operations which do no harm to the butterfly at all—we could prove that these tiny structures are extremely important in the grayling's mating, and presumably also in that of other meadowbrowns. To those who want to know more about it, I recommend reading the chapter which Niko Tinbergen devoted to the graylings in his recent book *Curious Naturalists*. At the same time, I hope that you will also read the other chapters in this fascinating volume. I admit that most of the dancing of the meadowbrowns takes place on the ground, but every once in a while we will see at least a beginning of it on, or near, flowers. This is the reason why I have brought up the subject here.

# 17

# On the Wings of the Night

The queen-of-the-night cactus, *Peniocereus Greggii,* is one of those plants that have helped to make Arizona famous. Yet, I have often wondered which would be easier, looking for the famous needle in the haystack or trying to find a wild queen-of-the-night plant in Arizona in the daytime. Really, there is no other cactus in the world which is so drab during most of the year. When you come across it for the first time—by sheer luck, of course—you will have a very hard time telling this dead-looking, sticklike thing apart from the dry vegetation around it. But don't be fooled! Just come back to the area on one of the few nights in May or June that this plant is in bloom. A pleasant, spicy fragrance now fills the atmosphere, and simply by following your nose you will soon come upon the flowers, which are of a breath-taking beauty. A single one may scent the air over a distance of 100 feet, but, for good measure, on a good and hot night some plants will display as many as 30 or 40 open blooms.

These flowers begin to unfold just after sundown, when the petals go through a series of jerky movements. Once open, the flowers measure from 4 to 6 inches in diameter. The inner "petals" are usually a creamy white, while the surrounding "sepals" may be lavender, green, brown, or purple. Although the blooms last through one night only, their ephemeral beauty is so impressive that this Arizona cactus is one of the most highly treasured of all night-blooming cactuses, of which there are quite a few.

Rather common as a cultivated cactus is the night-blooming cereus, which is not a cereus at all but a member of the genus *Epiphyllum.* In contrast to most other cactuses, which are as we all know desert plants, epiphyllums are native to the tropical jungles of Mexico and South America, where they grow high up on the bark of trees. The true night-blooming cerei or moon-cerei (*Selenicereus*) are climbing or trailing plants from Central and South America which produce on their vinelike stems the largest cactus flowers in the world, so incredibly beautiful that we have run out of adjectives to describe them. It is almost as if people have tried to establish a certain hierarchy for the degree of beauty of these flowers, with the result that, nowadays, we have a king-of-the-night (*S. grandiflorus* from the West Indies), a princess-of-the-night (*S. pteranthus* from southern Mexico and Central America), and—most beautiful of all—the true queen-of-the-night, *S. macdonaldiae*, from Uruguay and Argentina, whose white flowers with their golden sepals reach a size of no less than 13 inches. The length of the flower tube here is about 9 inches, a record which is shared with *E. phyllanthus*. However, many other night-blooming cactuses come very close in this respect (Fig. 69). As a matter of fact, when we compare all the various species we know in this group, it turns out that they have quite a few features in common. For instance, their odors are always strong and pleasant, so much so that people have likened them to just about all the pleasant fragrances they could dream up: vanilla, heliotrope, lily-of-the-valley, magnolia, lemon, violet, jasmine, hyacinth, gardenia, lavender, and lilac. Their colors are always light and delicate: pure white, creamy white, yellowish, light pink, or pale violet, with sometimes a contrast between a slightly darker "calyx" and a very light "corolla." At the bottom of the flower tube, just above the ovary, there is in all cases a good-sized or even formidable nectar chamber which may reach a size of one inch, for instance in *S. grandiflorus*. Nectar-guides, however, are lacking. The stigmas are curiously split, with spread-out lobes that are often cleft again, so that the end effect is that of an umbrella from which the cloth is gone.

What can be the meaning of this interesting assortment of features? A little reflection and a comparison with certain other flowers will give the answer. When I first heard of the Arizona queen-of-the-night, I immediately thought of a comparable western European plant, the honeysuckle *Lonicera caprifolium*. It, too, is famous for the spicy fragrance which its flowers emit in the evening. In the summer camps organized by Dr. Niko Tinbergen in

Figure 69. One of the queen-of-the-night cactuses.

Hulshorst, the Netherlands, we used to do interesting little experiments with these honeysuckle flowers. We would put them in a wooden box provided with a system of slits, in such a way that the flowers were invisible from the outside but their scent could disperse freely. At dusk, pine hawkmoths—which were very common in the area—would come to visit the box. They reacted to the flowers from distances of up to 10 yards. After some zigzagging and circling around the box, they soon found their way to the flowers

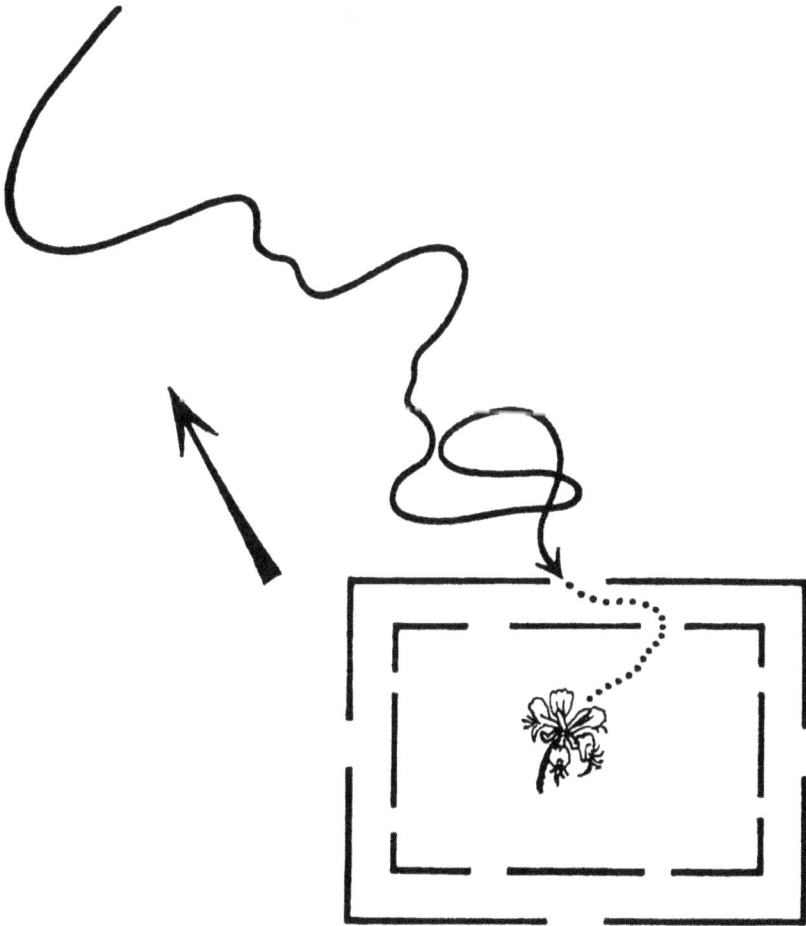

Figure 70. Experiment to demonstrate the attraction of pine hawkmoths by smell alone. Fragrant honeysuckle flowers are placed in the center of a perforated box which is covered by a second, similar box in such a way that visual signals are excluded. Wind direction is indicated by arrow and hawkmoth's path by tortuous line. (After N. Tinbergen, *Social Behavior in Animals*, by permission of the publishers, Methuen & Co. Ltd., London.)

inside (Fig. 70). Sipping their nectar, they would then pollinate them.

These experiments demonstrate in a very impressive way that smell alone can be enough to attract pollinating hawkmoths to a night-blooming plant. Of course, there are hawkmoths and hawkmoths. Some may be inclined to pay more attention to visual characteristics than others, and this may be especially true for species that fly in the daytime, such as our old friend *Macroglossa*. However, it is very likely that in all cases the fragrance will at least produce a food-seeking mood in the animals, making them eager to respond to color or, more likely, to light objects against a dark background—such as honeysuckle flowers.

If we are willing to believe that the flowers of night-blooming cactuses, like those of our European honeysuckle, are pollinated by hawkmoths, all the features we have mentioned become meaningful and everything falls into pattern. As we have already pointed out, in their feeding behavior hawkmoths are like hummingbirds to such an extent that they have often been confused with them. To get at the nectar, they hover before a flower and stick in their proboscis, which in some cases has a formidable length: a good 3 inches in the European bindweed hawkmoth, *Herse convolvuli,* and 10 inches in the tropical species, *Macrosilia cluentius.* The split-up, umbrella-like stigma of the night-blooming cactuses will catch a lot of pollen grains in various places, when it is touched by the soft proboscis of the hawkmoth.

It is very illuminating, in this connection, also to examine some other types of cactus flowers, for instance the rather shallow ones of *Pediocactus simpsoni* from eastern Washington. These blooms are open in the daytime, and it is quite obvious that they are pollinated by bees and bumblebees because they display the same trick which works so well for barberry flowers. Thus, when a bee in trying to reach the nectar at the base of the flower touches the slender stamens on the inside, they all begin to bend inward toward the pistil. Since there are so many stamens, it is just as if the insect is running the gauntlet, and the result is that it will be powdered all over with pollen. The pistil is so long that there is no risk that its stigma will be touched by pollen from the same flower.

After all the things I said above, we shall have no trouble in accepting, as good moth flowers, soapwort, *Phlox,* various silenes and wild orchids, jimsonweed, tobacco, and some evening-primroses. The common evening-primrose, *Oenothera biennis,* found along roads and in waste places both in Europe and the United States, is a very excellent example, but Americans are lucky in having some species which are even more spectacular. The yellow evening-primrose of the Midwestern plains, *O. missouriensis,* has flowers that measure 6 to 8 inches across. Almost equally large is *O. caespitosa,* a white-flowered species which ventures farther west across the Colorado plains and ranges throughout the Rockies. The blooms open in the evening, usually between 5 and 7. The whole process requires only 2 to 7 minutes, and it is pure delight to watch the thin white petals folding back and smoothing out their wrinkles. In most other flowers, the unfolding process is so slow that time-lapse photography is needed to get a good idea of it. A sweet fragrance soon fills the air, and if hawkmoths are in the area, they are not slow to respond. Some American hawkmoth species have tongues long enough to reach to the bottom of even the longest flower tubes. Darting about ferociously, they sip and sip until about an hour after dark, or until the supply is exhausted. Just as in the case of the night-blooming cactuses, the flowers last through one night only. The next morning you will find them drooping and wilted and turned into a lovely shade of pink.

To avoid giving the impression that all moth-flowers are of exactly the same type, with a flower tube as long as an elephant's trunk, let me discuss two other interesting cases. Figure 71 shows the flower of the martagon lily, or Turk's cap, native in central Europe but grown as a garden plant in many other parts of the world. In those places in the United States where the Turk's cap is hard to find, certain wild American lilies, which are very similar, can be substituted. The six petals are curved back upon the flower stem, so that the inch-long stamens stick out freely, together with the long and slender pistil. In the young flower, these stamens are straight, and the anthers form just an extension of the sticklike filaments. Later on, however, a graceful curvature in an outward direction takes place, and in addition each anther changes its position relative to its filament, so that the pollen compartments, which open by lengthwise slits, are now on the outside and down. This stage is represented in Figure 71. The anthers are attached by a small joint only, which gives them great freedom of movement;

Figure 71. Flower of Turk's cap lily, *Lilium Martagon*, a typical hawkmoth-flower. No perch is provided, but the visitors can easily circumnavigate this structure in their efforts to get at the nectar secreted by nectar grooves at the base of the petals.

this, in its turn, facilitates tremendously the dropping-out of the pollen. The same construction, by the way, is found in honeysuckle (Fig. 72). The whole plan of the Turk's cap seems to be aimed at making possible its circumnavigation by hawkmoths. Interested in the nectar which is secreted by the six long, tubelike nectar grooves found at the base of the petals, they make at least six stabs with their proboscis. In doing this, they just cannot avoid touching, on the pollen-liberating side, the anthers that swing to and fro so freely. The sticky pollen is later deposited on the excentric stigma of the same, or another, martagon lily flower. The normal pollinator is *Macroglossa*, whose inch-long proboscis is just about the right size to cover the distance between the pollen and the entrance to the nectar tubes.

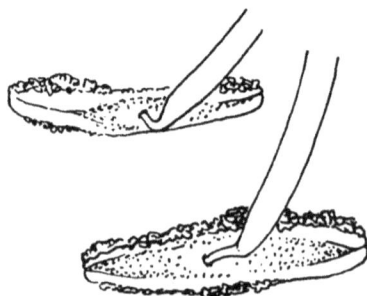

Figure 72. Attachment of the anthers to the filaments in the stamens of a hawkmoth-pollinated honeysuckle, *Lonicera*. The least little touch will cause the anthers to swing back and forth, releasing the pollen.

In contrast to what we have noticed in other hawkmoth flowers, the big, beautifully white, funnel-shaped blooms of the European bindweed, *Convolvulus sepium*, are practically odorless, at least to us. The most important visitor, and a very efficient pollinator, is the bindweed hawkmoth, *H.* (or *Protoparce*) *convolvuli*. Indeed, the geographical distribution in Europe of the hawkmoth and the bindweed is almost the same, that of the bindweed being a little larger. The dependence of the plant on the hawkmoth is not an absolute one, since it can also multiply rapidly without flowers, and can attract other pollinators such as bees, bumblebees, and Syrphid flies. Because of this, *C. sepium* does set seed in England, even though the pollinating hawkmoth is rare there. On the other hand, the claim has been made that it is not often found in Scotland, where the animal is completely absent.

The efficiency of *H. convolvuli* is incredible. Each animal seems to be in front of at least three flowers at the same time, so lightning-fast are its movements, and when a half dozen of these hawkmoths are at work on a field of bindweed plants, one gets the impression of dealing with an army of them. To give the reader an idea of the efficiency of hawkmoths in general, I present the following figures. Literally hundreds of primrose flowers or gentians can be pollinated by *Macroglossa* in a few minutes. One hundred and six flowers of *Viola calcarata*, a long-spurred violet of the Alps, were pollinated in less than 4 minutes, and 194 flowers in less than 7 minutes. Their rapidity of movement must serve the hawkmoths well when they are pursued by bats and also when they migrate; specimens of the oleander hawkmoth, which breeds in the Mediterranean region, have been found as far north as southern Sweden!

Quite apart from their behavior in pollination and migration, hawkmoths would still be fascinating creatures. Their caterpillars, easy to raise and so well known that they even have a common name (hornworms, for the curved, pointed "unicorn" on their posterior), often display a magnificent camouflage principle known as countershading. Here is how it works. Against a background of flat leaves, a sausage-shaped object such as a caterpillar tends to be conspicuous, in spite of the fact that it may have the same general color as those leaves. This is simply due to the fact that the round caterpillar is *shaded*. To cope with the situation, many caterpillars are darker on the side normally turned toward the light, with a very gradual change into the lighter hue on the opposite side. Under natural illumination, the result will be that the caterpillar appears in a perfectly uniform color; it looks just like a flat leaf. The roundness that would give the animal away to its enemies has miraculously vanished. The best way to find such

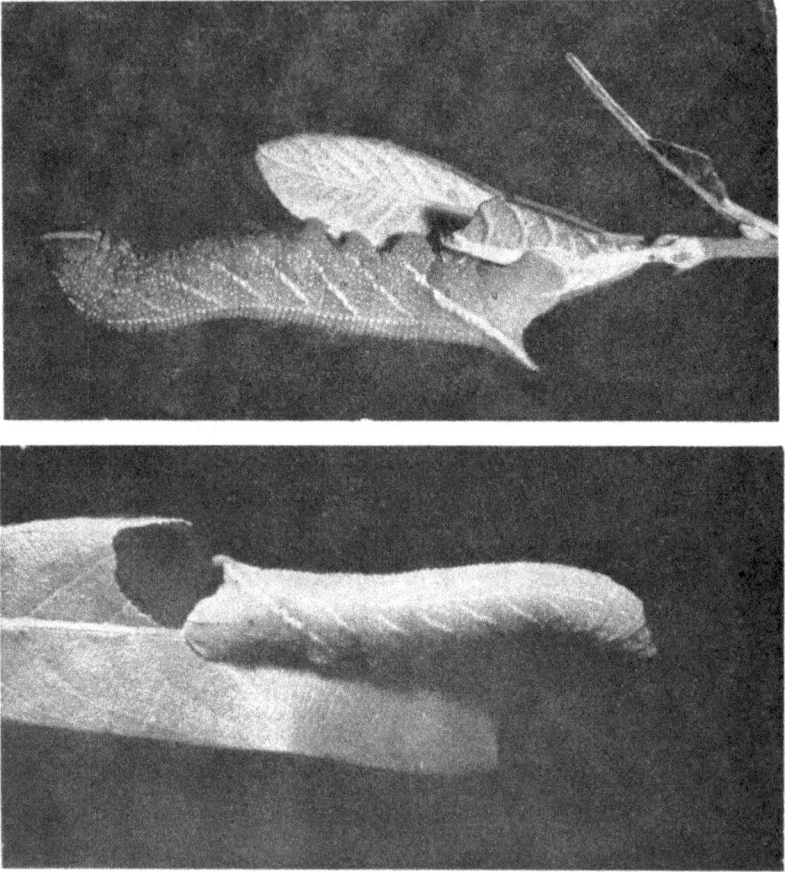

Figure 73. Hawkmoth caterpillar, showing the phenomenon of counter-shading.

caterpillars in nature is to turn the twigs of their food plants around. It is quite amusing to observe the look of amazement this will bring to the faces of unsuspecting people. Our photographs (Fig. 73) demonstrate the principle, using caterpillars of a hawk-moth from the Pacific Northwest as an example.

Some hawkmoths have played a role in mythology and folklore. The most conspicuous of these probably is the European death's-head sphinx moth, *Acherontia atropos*, which has dark markings suggesting a death's-head on its back and was considered an evil omen, announcing hunger and pestilence, in the Middle Ages. It has the habit of stealing honey from beehives, to which it is said to gain admittance by imitating the piping of the queen. At this sound, the worker bees become immobile, and the sphinx can get at the food stores with impunity. On each trip, it may sip up almost its own weight in honey, and since it is a big moth, this amounts to a small spoonful. It is often claimed that in regions where death's head sphinx moths are common, the bees barricade the hive's entrance with wax and propolis (a resinous glue which bees collect on young twigs) against the night, and there may well be a grain of truth in this story.

To get an idea about the hawkmoths common in the United States, I advise you to read Gene Stratton-Porter's delightful little book *Moths of the Limberlost*. The "bloody nose" moth which she mentions is a clearwing-hawkmoth, a creature not much larger than a bumblebee, to be found in brilliant sunshine visiting such flowers as thistles, vetches, and clover. As a matter of fact, there are several species of clearwing-hawkmoths in America, of which *Hemaris Thysbe* probably is the commonest. The rather unfriendly name "bloody nose" refers to the scientific name *Hemaris* which means the same thing; there is indeed some red color on the front end of the animal. However, if I had my way, I would call this animal the bumblebee-hawkmoth. It is indeed unbelievable what a magnificent job of imitating a bumblebee *Hemaris* does. It is not just a matter of having the right size and shape; the movement fits the bumblebee pattern, too. Whenever I recognize *Hemaris* in the field—and I am sure that I have overlooked many a specimen—I am thrilled. The almost perfect resemblance between the "model" (the bumblebee) and the imitator (*Hemaris*) is another illustration of the mimicry principle, already discussed in chapter 14, "The Imitators." Obviously, it is an advantage for a tasty animal such as a hawkmoth to look like an unpalatable bee, but in the particular case of *Hemaris* there is probably more to the story than just this. The similar way of life must have helped a little bit, too.

Another hawkmoth mentioned by Mrs. Stratton-Porter is the white-lined morning sphinx, *Celerio lineata*. She recalls how in her childhood days all her relatives mistook this moth for a hummingbird, flying as it did in the daytime to feed on petunias and other trumpet-shaped flowers. Figure 74 shows the animal in front of a *Rhododendron* bloom.

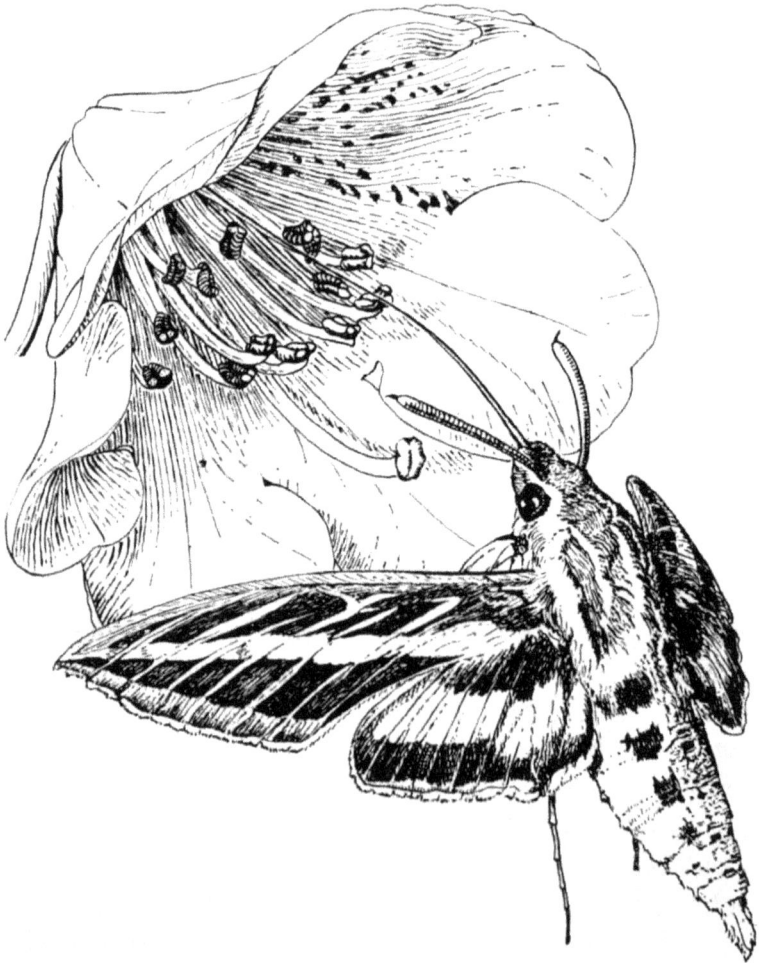

Figure 74. Hawkmoth (*Celerio lineata*) pollinating a rhododendron flower.

*Protoparce celeus*, the "king of the hollyhocks," comes forth at dusk and flies in the evening and at night. The caterpillar of this moth (Fig. 75) is one of the "tomato worms," big green fellows which can be found not only on tomatoes but also on other plants belonging to the nightshade family. The caterpillar pupates in the ground, changing into a chrysalis that is a delightful miniature of a Greek vase or jug (Fig. 76).

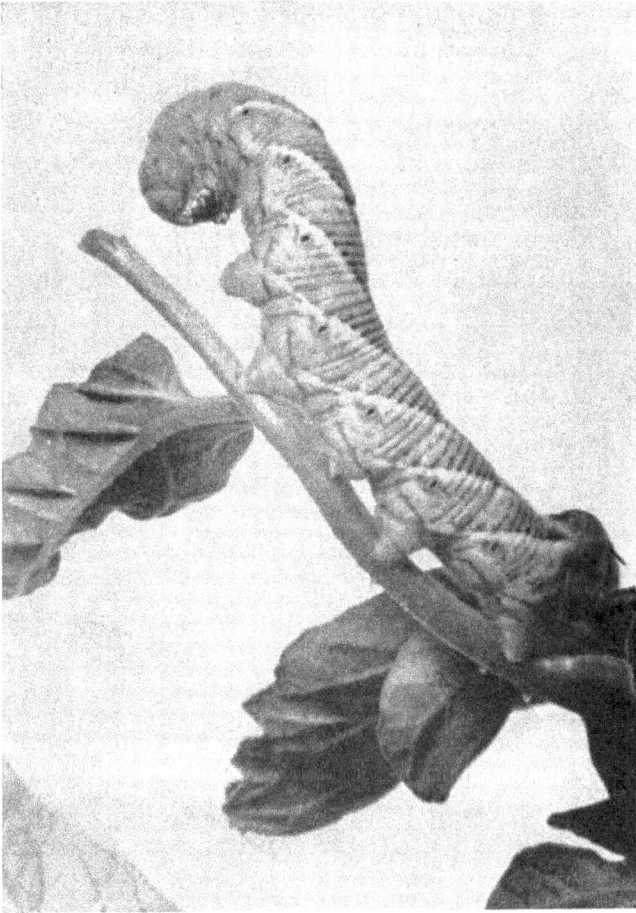

Figure 75. Tomato hornworm (larva of a tomato hawkmoth) in the sphinx position.

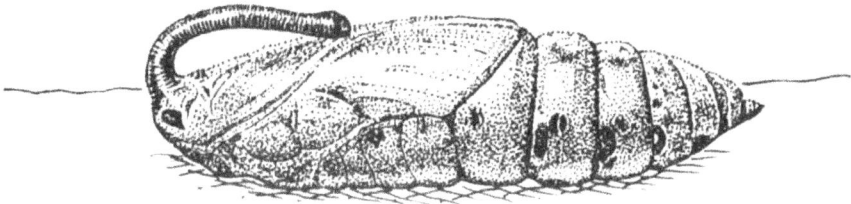

Figure 76. Chrysalis of the tomato hornworm, a tomato hawkmoth.

However, we do not try to create the impression that all the moths that are active in pollination are hawkmoths. There are many "owlet moths," members of the family Noctuidae, that do a good job, too. One of the most interesting ones, a moth that everybody in this country should know because it is so common and active, is the little gamma moth (*Plusia gamma* and some closely related species). It was named for the Greek character gamma, γ, displayed in a distinct, light color on the brownish-gray front wings. In Holland people call it the pistol moth because they think the figure is more like a pistol in the process of being fired. *Plusia* has a proboscis which is from 15 to 16 mm long, that is, more than half an inch, quite something for such a little animal. This instrument enables it to get the nectar from such "difficult," narrow-tubed flowers as red clover and that long-spurred stand-by of our gardens, red valerian or Jupiter's beard (*Centranthus ruber*). The activity of the gamma moths reaches its peak in the twilight, but I have often seen them working during the daytime and also on moonlit, starry nights until past midnight. *Plusia* has been used rather extensively in experiments on color vision. It can be trained to yellow and violet and must, therefore, see these as real colors.

*Mamestra* and *Dianthoecia* moths are good friends of the night-flowering silenes (catch-flies). However, for their good services in pollination they get more in return than just nectar; the females deposit their eggs in some of the flowers, and the young caterpillars live at the expense of the pistils and the ovules therein. This reminds me of a case that I can only describe as an absolute masterpiece; let me present it here as the climax to my story.

A complete, unbreakable, and unshakable tie-up between plant and pollinating insect is found in the case of the Spanish dagger plant (*Yucca*) of Mexico and our southwestern states and a small moth, *Pronuba yuccasella*. Although most yuccas can and do reproduce vegetatively, the moth at least could not survive without the plant. The reason is that the small caterpillars of the moth grow up in the lower part of the pistils, the ovaries, of the *Yucca* flowers. Here they find food and shelter. On the average, each caterpillar eats up about 20 of the developing seeds, so that one could rightfully claim that they are harmful. On the other hand, since each flower contains about 200 ovules and does not usually give shelter to more than six caterpillars, there will always be some good seeds left. Also, and this is much more important, there would not be any seeds in the first place if it were not for the activities of the moth.

At the time when the *Yucca* flowers open, the female *Yucca* moths are already on their way to them. In the light of the moon, they alight and penetrate into the flowers, and each begins to form a neat, compact, little package out of the pollen grains which it finds. It turns out that the shape of the front legs and the mouth parts, and also the way in which they operate, are just right for this task; in fact, it is just about the only thing that can be done usefully with these organs. With her little pollen package "under her chin," each female *Yucca* moth now flies to another flower, where she will deposit some eggs in the ovary. After that, she climbs up along the style and deposits the package in the deep hole that forms the upper part of the stigma. The condition of the stigma being what it is, this is the only way in which pollination and subsequent fertilization of the ovules can be brought about. Without the moth, the *Yucca* plant would not be able to reproduce sexually—a distinct disadvantage in the struggle for life. Conversely, the moth would soon become extinct without the *Yucca* plant. The amazing thing is that the moth behaves like an intelligent being, and yet she cannot possibly have any real insight into the things she is doing. She does not understand that the deposition of the pollen package in the stigma leads to the formation of good seeds. Since her life span is short, the results of her acts will never be known to her, so that experience cannot have been her master, either.

# 18

# The Blundering Beetles

When we were little boys, our favorite ponds in the Buitenzorg Botanical Gardens were the ones in front of the Governor-General's palace. Here one could see black Australian swans sailing by majestically, true oriental lotus plants raising their "sacred" flowers out of the water, and also a number of huge round leaves with up-turned rim, floating on the surface. With a certain awe, we would tell each other that these were the leaves of the famous *Victoria regia* * plant from the Amazon region in Brazil, and that a small child could safely sit on them. (I am sorry to say, though, that our own trials were not very successful!) Occasionally, there would be blooms. We knew that the flower buds would open in the evening and that the beautiful, creamy-white flowers would then be warm, emitting the sweet, mysterious fragrance of tropical fruit. The next day we would find these same flowers closed again. They would once more open in the evening but now the color would be purplish and the odor almost gone.

In those days, the behavior of *V. regia* did not make much sense to us, but it does indeed have its purpose. In its native environment in Brazil, the fragrant fresh flowers will attract large numbers of beetles, related to our June bugs and Japanese beetles. During the night, when the *V. regia* flowers gradually "fold up," these insects will become prisoners. Very destructive prisoners, too, who feed on the juicy inner wall tissues of their chamber. However, at the same time they may bring about pollination, for as a rule they have brought with them pollen from a previous visit to a *V. regia* flower, and the stigmas in the fragrant prison are at just the right

* The correct botanical name is *Victoria amazonica* (Poeppig) Sowerby.

stage to receive it. The next evening, just about at the time when the flowers unfold again, fresh pollen is shed. The beetles, covered with the powder and no longer held spellbound by sweet odors, take to the wing to drop in on another fresh and fragrant *Victoria* bloom.

It is too bad that we did not suspect the role of beetles in our *Victoria,* for we already had an inkling, at least, of the relations between beetles and another extremely remarkable plant: *Amorphophallus variabilis* or snakeleaf, big brother of the little arum lilies we described in Chapter 13. On the island of Java, *A. variabilis* was both common and well known. People referred to it as kembang bangke or carrion flower, because the inflorescences have a smell that is simply terrible. It is not at all amazing that they attract carrion beetles in large numbers. These animals will stay for no less than five days, in spite of the fact that there is no real trap here as in the case of the spotted arum. The reason is that the plant offers the beetles, at the bottom of the spathe, an edible layer so delicious—to them!—that there simply is no advantage in leaving. The female flowers that are to be found in this region happen to be receptive at this stage, and there is a good chance that at least some of the beetles have come in with pollen to fertilize them. By the time the delicious layer begins to rot, toward the end of the five-day period, a pollen rain from the male flowers starts to fall down on the beetles, so that when they sally forth on their quest for another inflorescence they are simply loaded with the flower dust, and the story can be repeated.

Is it fair to bring in these examples from the tropics? Nowadays, the answer must be Yes. A close relative of the kembang bangke, *A. titanum* from the jungles of Sumatra, bloomed in the famous botanical gardens at Kew, England, in 1926, and in this country at the New York Botanical Gardens in 1937 and again in 1939. It has also been in flower in Hamburg, Germany, and Wageningen, Holland. The inflorescence, in this case, may be 8 to 10 feet high (Fig. 77). It produces a considerable heat, and the carrion smell is so overpowering that some people are known to have fainted when they came too close. In nature, the pollinating carrion beetles are held prisoner in a flower chamber from which escape is made impossible, not by groups of hairs or bristles but by a very sharp edge or keel on the central column.

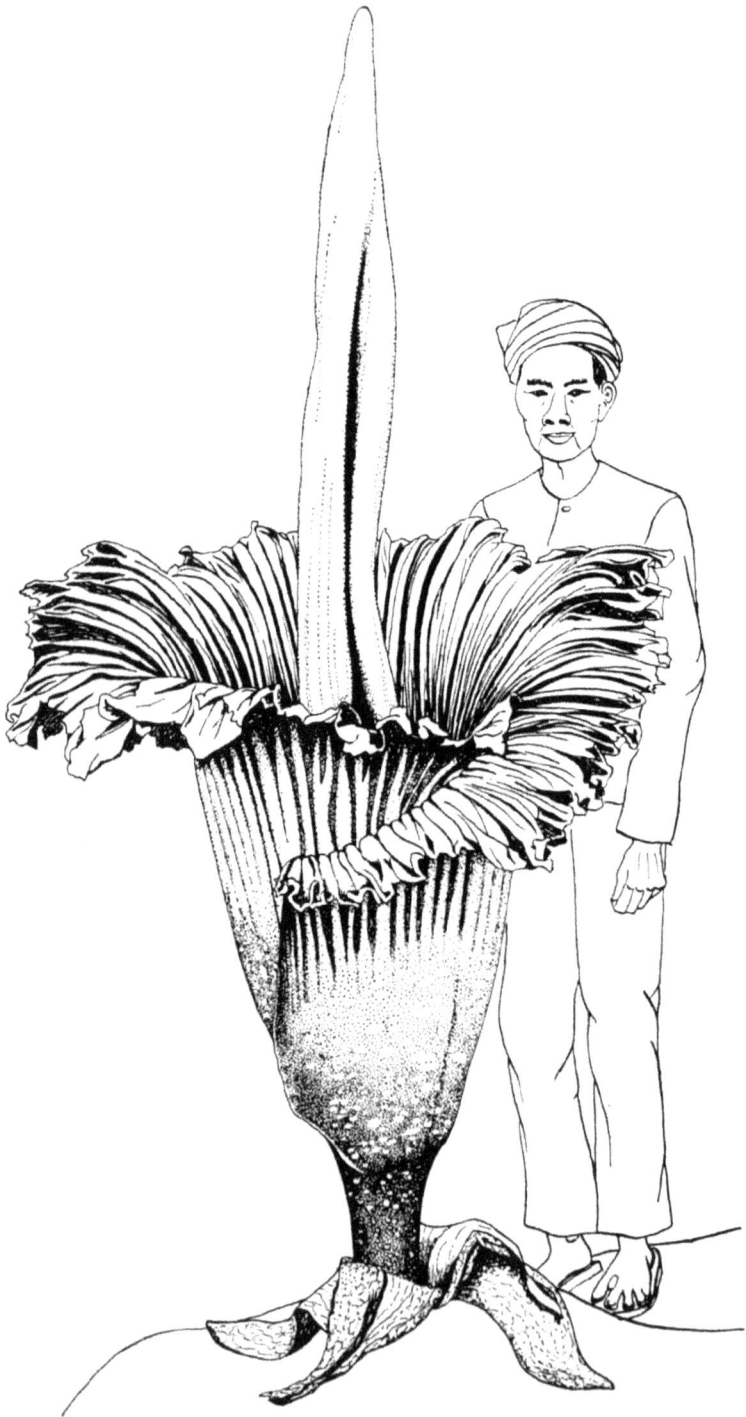

Figure 77. *Amorphophallus titanum*, a giant arum lily from the jungles of Sumatra. It produces an overpowering carrion smell and is pollinated by carrion beetles.

A case of beetle-pollination in arum lilies with which American city-dwellers can familiarize themselves even more easily is that of *Dracunculus vulgaris,* known in some circles as dragon lily, black calla, or black lily of the Nile. Although it is found naturally only in the northern Mediterranean region, from Portugal to the western part of Asia Minor, this plant has found its way into many American gardens. It is very popular and thrives in Seattle, where it flowers around the middle of June. *Dracunculus vulgaris* and its variety *creticus* from the island of Crete possess the largest flowers, or rather inflorescences, in the European flora. The average size is from 1 to 2 feet, but inflorescences that are 3 feet tall are by no means rare. The inside of the spathe has a beautiful, purplish-black color. The appendix is even darker, having the glossy appearance of black cherries.

The spathes of *Dracunculus* open in the morning hours. Although the heat development is not impressive—the reason for this being that the appendix is hollow and thin-walled, so that there just is not enough fuel to burn—the smell, which reminds one of long-dead fish, is overpowering. The air is soon abuzz with flies attracted by the appendix, and some of them are fooled by its odor to such an extent that they will deposit their eggs on this organ. Most of them do not enter the flower chamber. Small beetles, on the other hand, can soon be found in the chamber by the dozen. On a sunny afternoon in Seattle in June, 1959, I collected 162 of them from one single inflorescence! From two *Dracunculus* plants, I collected a total of 298 beetles, belonging to 15 different species. As could be expected, they were of the type usually found on excrement and carrion; only 2 of them were compost beetles.

If the inflorescences are inspected on the second day of flowering, the beetle harvest is poor. Most visitors have already left, after having been showered with pollen, and the fish smell is gone. We can only marvel at the wonderful timing of the events that take place in *Dracunculus* and its relatives. Some of these can indeed be used as clock flowers, since they always begin to smell at a certain hour. For example, the kembang bangke of my childhood days would always do it at 4:30 P.M. My colleague Van der Pijl, who like Knoll has done excellent work on arum lily "flowers," has found how the chain of events is triggered in this case. It is the change from darkness to light, at about 6:00 A.M. in tropical countries, which starts a chemical process that results in the production of stench 10½ hours later. Thus, to obtain plants that begin to smell at, say, 3:00 P.M., it is sufficient to put them in the light a few hours before daybreak. Likewise, if it is desired to start them at 7:00 P.M., they should be kept in darkness until about 9:00 A.M.

Van Herk, who has written a beautiful thesis on the biochemical events in the so-called "voodoo lily" (*Sauromatum guttatum* from northwestern India and Pakistan, another arum lily), has demonstrated that in this plant the process in which heat and stench are produced is started by a hormone, calorigen, originating in the male flower buds. It begins to leave these buds about a day before the inflorescences unfold. No heating will therefore occur if the male buds are removed one day or more before the expected opening time. However, if the critical moment has passed, the avalanche has started and removal cannot change the picture at all.

It is almost superfluous to point out that there is a beautiful correspondence between the behavior of the inflorescence and the habits of the natural pollinators. Many carrion beetles reach a peak in their activities in the afternoon and early evening, and it is therefore quite meaningful for the kembang bangke to begin its smell production at 4:30 P.M. To those of you who seek a North American example, I could mention spicebush or *Calycanthus*. Not too long ago, Verne Grant made a beautiful study of the flowers and the pollination in the western species, *C. occidentalis*, a shrub which is not uncommon along streams and in shaded canyons in northern California. From May to September, the plants produce an abundance of solitary flowers, fairly big and dark red when they open. Just as in *V. regia*, we find a strong odor at that time, but in this case it is the heady fragrance of wine. In its structure, the *Calycanthus* flower can be compared to a cup (Fig. 78). It has a large number of free "pistils" on its inner, hollow surface and a continuous series of bracts, petals, stamens, and staminodia, or sterile stamens, on its outer surface and upper rim.

When the stigmas are ready to receive pollen, the outer petals will spread out, but the inner ones do not follow suit; instead, they form a central cone with a very narrow opening on top, more or less like a crater. This provides an excellent vent for the odor, and indeed it does not take long for the flower to lure into the dark crater depths a number of tiny beetles, about 3 mm long, belonging to the family of the flower beetles, or Nitidulidae. One may find

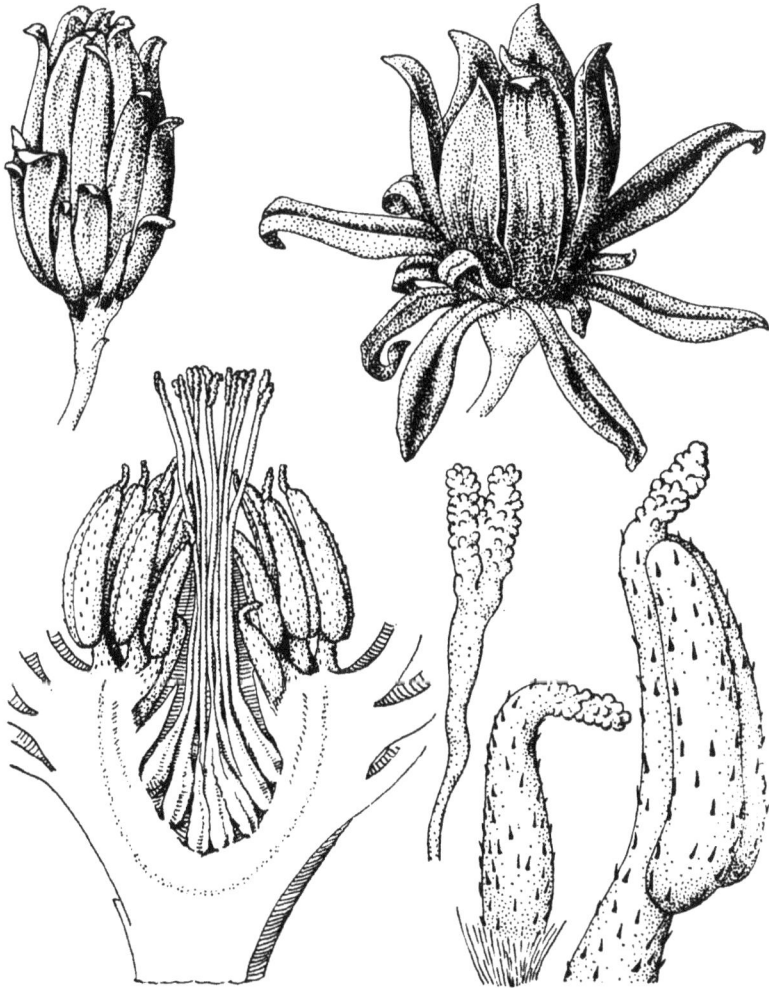

Figure 78. *Calycanthus,* a beetle-pollinated flower. Upper left: young stage. Upper right: older stage with formation of a crater serving as a vent for the odor. Lower left: lengthwise section through center of flower to show arrangement of floral parts. Lower right: upper part of pistil, staminodium and stamen. Notice the nutritious, cauliflower-like tissue in all three and also the bristles at the base of the staminodium.

as many as eight or ten of these *Colopterus truncatus* beetles in one flower.

Getting in is easy for them, as the inner surface of the petals that form the crater is covered with small bristly hairs which point downward. The beetles can also crawl upward on the staminodia

and stamens, which bear similar bristles pointing skyward. However, getting out is another matter, and getting down to the very
base of the pistils, where the beetles might do a great deal of
damage to the ovules, is likewise made very hard or impossible by
a dense growth of long hairs at the base of the staminodia. So, the
insects are kept trapped in the flowers at a certain level. However,
they are well fed; the upper part of the innermost petals, the
stamens and staminodia, and even the receptive tip of each stigma
consist of a pearly-white, nutritive tissue. There is no reason, then,
why the tiny *Colopterus* beetles should not feel perfectly happy.

After from 12 to 36 hours, probably depending somewhat on the
weather, the anthers will open and shed their pollen, dusting the
beetles. The stigmas, however, are protected against self-pollination because the inner staminodia, in the meantime, have bent inward, placing themselves like so many shields over the stigmas.
Still, a few hours later, the cone-forming petals will bend outward
a little way, so that the flower is now more open. The succulent
tissue withers, the fragrance disappears almost completely, and the
petals fade and turn brown. The *Colopterus* beetles, no longer interested, take to their strong wings to repeat their performance in
another *Calycanthus* flower.

Verne Grant considers it very likely that in the spicebush fertilization does not occur unless some animal carries pollen from
one flower to the stigmas of another. Bees and flies do not seem
to pay attention to *Calycanthus* flowers and, incidentally, would
usually be too big to enter, anyway. Thus, the spicebush flower is
entirely dependent on *Colopterus* or other small beetles for its pollination. *Colopterus*, on the other hand, has been reported from
various localities between Canada and Brazil, where spicebush is
not always found, and must therefore be considered able to fend for
himself.

Other beetle flowers in the American flora are magnolias and
peonies and at least some of the water lilies. There are many
people who believe that these are "primitive" flowering plants, that
is, they are thought still to be very much like the ancient relatives
of present-day plants; most of the latter have changed into more
"modern" types during the process of evolution. It is interesting
that it should be just exactly this "ancient" group of water lilies,
magnolias, and peonies, the Ranales, that has connections with

beetles. These insects, too, are considered to be "old" or "primitive." They were already highly developed in the Mesozoic geological period, when they must have formed more than a third of the total insect fauna. More than 400 fossil, Mesozoic beetle species have thus far been described! In those days, there were only a few flies and beelike insects and no butterflies or moths whatsoever. These and other flower-lovers, such as true bees and wasps, beeflies and Syrphid flies, did not really get their start until the early Tertiary period. In other words, in the distant days when the flowering plants had just begun to appear, beetles were practically the only insects that were around to enter into any relationship with them.

Perhaps it is not a coincidence that there is at least one beetle-pollinated cycad, too, *Encephalartos villosus*. It is well known that cycads, somewhat palmlike in appearance, are primitive plants. They are, in fact, even more primitive than the conifers (spruce, pine, and the like) and the Ranales which we just mentioned. Verne Grant believes that the Mesozoic relative of the cycads, *Cycadoidea*, was also beetle-pollinated. According to him, it had several features in common with our present-day *Calycanthus*, such as the presence of a traplike structure in the inflorescence and the fact that the female organs reached maturity before the stamens did.

Be that as it may, the evolutionary performance of the group of the beetles has certainly been very disappointing when things are considered from the point of view of the flowers. Although nowadays there are at least 300,000 beetle species—and we are discovering new ones every day!—only a very few of these have developed mouth parts resembling those of bees and butterflies. Especially fascinating among these exceptions are the species of the genus *Nemognatha* in the family of the blister beetles. *Nemognatha* (literally: threadjaw) occurs in tropical and subtropical America, with several species in Oregon and Washington. The maxillae or lower jaws are developed to such an extent that in many species they are longer than the body, or, rather, each maxilla bears an almost straight, saber-like appendage (Fig. 79). The inner, hair-covered margins of the "sabers" touch each other, so that a sort of gutter is formed through which nectar can move to the beetle's mouth, with the aid of capillarity. The gutter can with

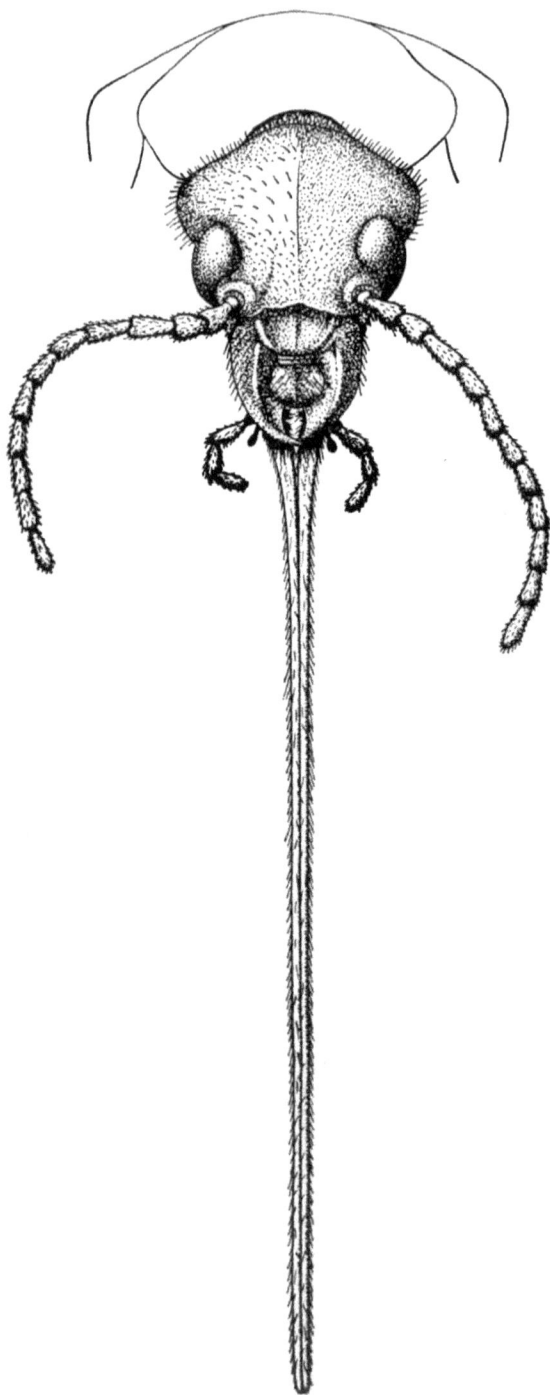

Figure 79. Head of a beetle belonging to the genus *Nemognatha*. This is one of the extremely rare cases where the mouth parts of a beetle are adapted to long-tubed flowers.

good reason be compared with the proboscis of a butterfly, for the mouth parts that form the structure are the same in both cases. However, the instrument of *Nemognatha* cannot be coiled and uncoiled, like that of the butterfly, and of course the mechanism of "sucking" is different, too.

Most other beetles have two pairs of "biting jaws" and a short lower lip, so that they cannot get the nectar from deep, tubular flowers. The best they can do is to crush pollen grains with their jaws and lap up exposed nectar with the hairy parts of their mouths. The places to look for flower-visiting beetles, therefore, are the easily accessible inflorescences of such plants as wild carrot, angelica, elderberry, and cow-parsnip. Wild rose flowers are very good, too; and, although the American chestnut has in many places been wiped out by blight, I should also mention the inflorescences of these trees, for in central Europe they show at least an approximation to predominant beetle-pollination. The male flowers of the chestnut have a strong, somewhat fishy smell. They will, therefore, attract all kinds of insects: bees, wasps, flies, and butterflies as well as beetles. Yet, there are places where more than half of the visiting insect species are beetles, for instance, 53 out of a total of 103 in Carinthia, according to Porsch. The staminate and the pistillate flowers are found close together, and both nectar and pollen are offered in profusion. As long as it is not too old, the pollen is sticky, which favors transport by insects. However, toward the end of the flowering period, the pollen loses its stickiness, so that now the wind can begin to play a role.

# 19

# Of Sausage Trees and Bats

. The town of Bogor, on the island of Java, has long been famous for its Botanical Garden. Even in the yards of private citizens one can now find some most remarkable plants and trees (which must have gotten there in some funny way, no doubt!). I remember vividly how impressed I was, as a little boy, every time I saw a sausage tree in fruit in the garden of my friend Rooster. As its scientific name *Kigelia aethiopica* indicates, this tree comes from Africa, but what awed us even more were the fruits: big brown sausages, suspended in mid-air by means of slender, vertical, rope-like branches (Fig. 80). One could not eat these fruits because they were as hard as brick. And the flowers that were there before the sausages appeared were very unusual, too, showing a murky purplish-green color that was a far cry indeed from the attractive, bright hues associated with bee-flowers. Their smell reminded us of the cage in which we kept our white mice. To get a whiff of that musty odor one had to be there after dark; not too early, because then all the flowers would still be in bud, but not too late either because early in the morning they would invariably be lying on the ground, spent. And what eerie noises one could hear by staying close to the open flowers for a while! On several counts, then, that sausage tree was the strangest plant in our environment.

I know now that the strange noises were made by bats. Under a flowering sausage tree, they fly back and forth at breakneck speed, sitting down briefly, now and then, on one of the suspended flowers to lap up some nectar. Fighting and screeching occurs whenever two bats try to get at one flower at the same time. These bats take care of the pollination of the sausage tree (Fig. 81). Some of them are real specialists, for whom nectar and pollen are the only source of food. Their head is long and narrow, so that they can stick it into flowers without trouble. Whereas they lack teeth almost entirely—the few they have are used only for defense—

Figure 80. Sausage tree (*Kigelia aethiopica*) in fruit. This tree is pollinated by bats. The flowers, and later on the fruits, are suspended by means of rope-like branches, facilitating the approach of the bats.

they do possess a long tongue that can be stuck out very far. One of the Javanese flower bats, in fact, has the scientific name *Macroglossus* which means big-tongue! The tip of the tongue is covered with a number of soft, long "warts," pointing toward the base. This undoubtedly helps in the taking-up of the nectar, which in all probability is not a sucking process but a lightning-swift sticking-out and drawing-in of the tongue. In the special case of *Macro-*

*glossus*, that organ is shaped like a brush. It is not too amazing that we find almost exactly the same structure in certain parrots which also feed almost exclusively on nectar and pollen.

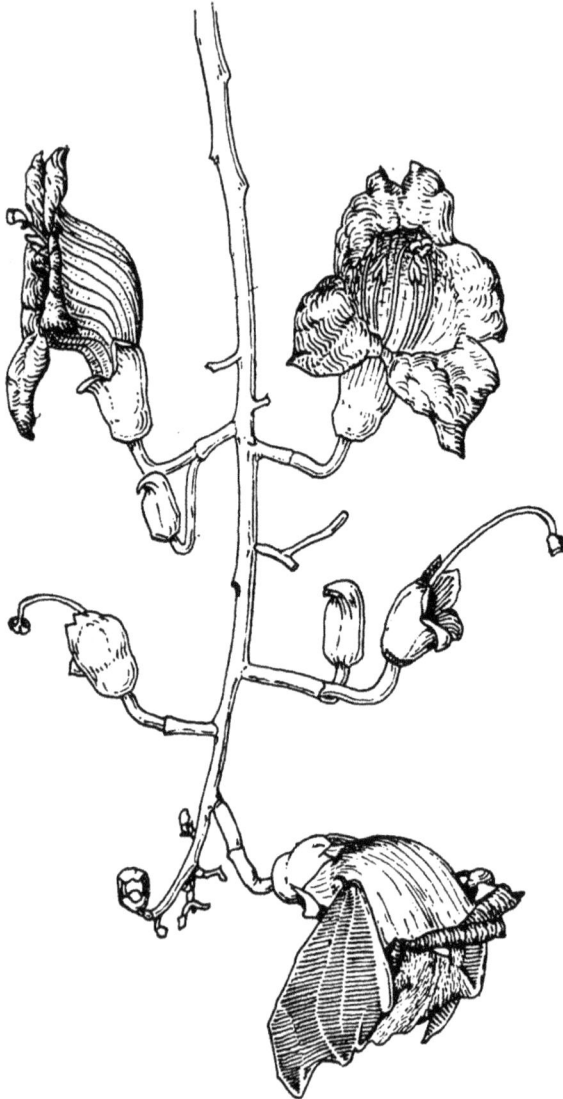

Figure 81. Bat pollinating a flower of the sausage tree, *Kigelia aethiopica.*

As we shall presently see, many tropical plants and trees are pollinated by bats. It is an interesting fact that the nectar specialists among these animals are the offspring of two different groups. In Central and South America, we find a few descendants of insect-eating bats, Microchiroptera, that have become flower-lovers. On the other hand, the ancestors of the African and Asiatic flower bats were fruit-eaters, belonging to the Macrochiroptera. It is still possible to find what we might call "transitional stages" at the present time. A truly classic example of these is offered by *Freycinetia insignis*, a Javanese liana related to the well-known pandanus trees of tropical shores. W. Burck, who was one of the directors of the Botanical Garden in Bogor, discovered in 1892 that the big fruit-eating bats known as "flying foxes" (*Cynopterus*) would, in the evening, visit the freycinetias to feast on the juicy bracts surrounding the inflorescence. Later investigators have expressed some doubt with regard to Burck's observations. The origin of their skepticism was the fact that there are other freycinetias which are clearly bird-pollinated, producing red and odorless flowers that are open in the daytime. Recently, however, Van der Pijl has been able to confirm in a beautiful way Burck's vaguely worded report, pointing out that *F. insignis* has pinkish to pale-lilac flowers that will open and spread their musty odor in the evening. The fruit-eating flying foxes that pollinate them are treated to a meal of juicy inner bracts and are offered special, erect, solid food bodies as well. Just in passing, we would like to mention that in New Zealand the white or pale-lilac bracts of *F. Banksii* are also eaten by man.

We have already mentioned some of the characteristics of flowers that are bat-pollinated. They always have a very peculiar smell. H. G. Derx, in the East Indies, has been able to isolate, from a *Fagraea* species, one of their odoriferous substances in pure form. It turns out to be that remarkable compound diacetyl, which in very small quantities can be used to impart a butter flavor to margarine but which has a repulsive, mouselike odor when it is concentrated. Bats seem to love it in any form.

The position of bat-flowers is such that the bats can get at them easily. If the flowers are not suspended from ropes, like those of our sausage tree, they are attached to the tree trunk, or they stick out freely above the crown. In this way, the bats can never get hurt by tangled branches. Not a superfluous precaution! True, A. Lazaro Spallanzani in the eighteenth century was already familiar with the fact that certain bats can easily find their way around in the dark, and we know nowadays that they employ very successfully a principle similar to our radar or sonar. But it is a curious fact that the flower-loving bats are much less dexterous in avoiding obstacles than their insect-eating cousins; their "radar sense" must be poorly developed. On the island of Amboina, where bats are eaten by man, it is indeed common knowledge that the flower-loving bats can be caught in nets with much greater ease than their relatives.

Bat-flowers or inflorescences are so strong that they are able to bear the weight of a visiting bat for a while. Often, the latter will "hook himself in" with the thumb-claws of his wings. Since one flower may be visited by a large number of bats in one evening, it is not unusual to find flowers with several dozens of the telltale little holes in them the next day. The picture is so characteristic, that nowadays it is considered to be good evidence for bat-pollination, even in those plants where the actual visits of the animals have never been observed.

The mouth of a typical bat-flower is very wide, of course, for there must be easy access to the nectar. This fluid is produced in great abundance, and so is the pollen. The colors of the flowers are not very conspicuous, which again makes good sense.

As further examples of bat-pollinated plants of the tropics we could mention calabash (*Crescentia cujete*), balsa (*Ochroma lagopus*), kapok (*Ceiba pentandra*), durian (*Duri⌐ zibethinus*), several of the bananas (*Musa* species), and many others. However, what chance would an inhabitant of Europe or North America have to observe the thrilling phenomenon of bat-pollination with his own eyes? I am sorry to say that his prospects are dim. To be sure, sausage trees are found in Hawaii, the fiftieth state of the United States, but they are not native there and the pollinating bats are lacking. When, in spite of this, we notice some sausages on the Hawaiian trees, we can be sure that they owe their existence to artificial pollination and to the occasional activities of hawkmoths. In Africa, hawkmoths have been caught, and even photographed, in the act of pollinating the native kigelias.

Now that Hawaii is out, we must turn our attention to Arizona. Gratifyingly, the extreme southern part of this state has two species of nectar bats, *Choeronycteris mexicana* (Fig. 82) and *Leptonycteris nivalis*. Admittedly, nobody has ever observed them in action on Arizona flowers, but it is very likely that they are responsible for the pollination of the giant saguaro cactus. Otto Porsch, that grand old man of pollination, has pointed out that the saguaro flowers have all the characteristics of typical bat-flowers. It is known that *Choeronycteris* and *Leptonycteris* migrate to Mexico, and this is as it should be because, like hummingbirds, they are completely dependent on their food plants. If the flowering periods of these do not overlap, that is, if there is a flowerless period or season, migration to more hospitable places is the only way out unless the bats resort to hibernation (or aestivation) in caves. Much too little is known about all these matters, and an interesting field of study is thus still available.

Figure 82. Side view of the head of a Mexican flower bat (*Choeronycteris mexicana*), showing the paintbrush-shaped tongue. (After Dobson.)

Yet, it can be claimed that every inhabitant of the United States can get at least a glimpse, an idea, of bat-pollination. *Cobaea scandens*, the well-known cup-and-saucer vine which must have originated in Mexico, can be grown outdoors in many places in the United States, and is raised as a greenhouse plant in others. In Europe, where the plant was introduced in the eighteenth century, people have ventured all kinds of suggestions about its pollination, without ever coming up with a satisfactory answer. In 1958, at long last, S. Vogel, a scientist who did some very nice work in Central America, presented incontrovertible evidence for bat-pollination in a *Cobaea* species so closely related to *C. scandens* that it may well be its ancestor. It is very tempting, therefore, to postulate bat-pollination for *C. scandens* flowers, too. In the light of this hypothesis, all their peculiar features would be meaningful: the wide bell shape, the color, the strange odor, and (last but not least) the position of the flowers at the end of slender, naked, foot-long branches (Fig. 83). On a minor scale, this is the same phenomenon of "flagelliflory" which we so marveled about in the sausage trees, with their long ropes. We have to concede, however, that the last word on *C. scandens* and its possible bat-pollination has not been spoken yet.

Figure 83. The well-known cup-and-saucer vine, *Cobaea scandens*. Grown extensively in European and North American gardens and greenhouses, it is the immediate descendant of a bat-pollinated plant from Central America. Notice the long, naked twig bearing the flower.

# 20

# *Where Animals Are Not Needed*

Thus far in this book we have been singing the praises of all the various animals that are active in pollination. Both as a matter of honesty and for reasons of completeness, it is indeed high time that we consider those cases where the job can be done without them.

First of all, in spite of Darwin's famous dictum ("Nature tells us in the most emphatic way that she abhors perpetual self-fertilization"), it is a fact that successful self-fertilization occurs with great regularity in a large number of plants, including important agricultural species such as peas. Then there is pollination by the wind in grasses, evergreens, and various spring-flowering deciduous trees. In a few cases, water acts as the helper, and finally there are some very exceptional cases where no pollination of any sort is required.

## SELF-POLLINATION

In the year 1891, Anton Kerner von Marilaun listed no less than 20 different ways in which flowers can make sure that self-pollination occurs when cross-pollination is not forthcoming. Undoubtedly, he was a little overenthusiastic, overlooking or neglecting the fact that self-pollination does not always lead to fertilization and the development of good seed. It is therefore wise to do some picking and choosing before quoting Kerner's examples. Nevertheless, some of them are valid.

In foxglove, the whole corolla, including the attached stamens, is shed at the end of the flowering period, but in the process the anthers may very well touch the stigma and achieve pollination if

209

it has not taken place earlier. In still other cases, the style will eventually curve or go through all sorts of contortions so that the stigma lobes may touch the anthers of the same flower. Fireweed is a good example, even though we must concede that its flowers seem primarily designed for cross-pollination, being proteranderous. Earlier, we have mentioned Sprengel's observations on this matter.

In addition to these examples, there are flowers that are always self-pollinated. Extremely interesting among these are the cleistogamous flowers. By cleistogamy we mean the production of good fruits and seeds by flowers that never open. It was discovered as early as 1732 by Johann Jakob Dillenius, who seems to have used it as an argument against the sexuality of plants. He was highly critical of Sébastien Vaillant, one of the early advocates of the idea of sexuality, and as unprejudiced arbiters we must agree that Vaillant, although essentially right, was rather crude in his comparisons and not always fortunate in his wording. The plant that acted as Dillenius' guinea pig was rightfully called *Ruellia clandestina* by the great Linnaeus. In the year 1753, the latter was already acquainted with at least a dozen different plants that showed the phenomenon; a great many more have been added since.

Conspicuous by their absence on Linnaeus' list are the various cleistogamic violets, for nowadays it is they that we think of first when we discuss the phenomenon. *Viola canina* is a good example. It produces two types of flowers: big normal ones and smaller ones that remain closed. In their structure, the big flowers are very much like the ones we described earlier for *V. odorata*. They are, therefore, completely dependent on insect visitors and will simply die when cross-pollination is not forthcoming, as we indicated or at least implied in Chapter 16, "Flashing Beauties and Dashing Flyers." The flowers that remain closed owe their smallness to the fact that all the "non-essential" organs in them have been suppressed. Thus, the calyx lobes are only half the normal size, and the corolla lobes appear as minute scales. There are only two stamens in these flowers, instead of five, with only two pollen sacs each instead of four. However, the few pollen grains that are formed are quite normal, and so is the ovary. Germination of the pollen grains takes place while they are still in the anther. To reach the pistil, the pollen tubes only have to pierce the wall of the anther that separates them from the adjacent stigma. The fruits formed in these cleistogamic flowers not only are normal, but also seem to mature a little faster than those from the big flowers. The seeds, likewise, are quite normal and viable, so that in this case certainly Darwin's dictum does not apply.

## WIND-POLLINATION

It is generally conceded that the gymnosperms, to which such plants as firs, pines, cedars, and cycads belong, are more primitive than the flowering plants. Since, as a rule, the gymnosperms are pollinated by the wind, many people like to think that wind-pollination is always a primitive thing—even if it occurs in flowering plants. But is it? Let us consider the evidence in an unbiased way.

First of all, we notice that there are at least a few gymnosperms that are insect-pollinated. In Chapter 18, "The Blundering Beetles," we mentioned *Encephalartos villosus,* and we can now add *Ephedra campylopoda,* a plant for which Porsch demonstrated insect-pollination as long ago as 1910. We have also seen that Verne Grant believes that *Cycadoidea,* a possible ancestor of the flowering plants, was beetle-pollinated. All this leads us to believe that, in the course of evolution, a very gradual change took place in certain gymnosperms so that they became insect-pollinated, and that perhaps all the early flowering plants that arose as their descendants were, as a consequence, also insect-pollinated.

The change in character of the pollen required in this evolutionary process was quite profound. A wind-borne pollen grain must of course be light, but above all dry and not sticky; otherwise, it would tend to form clumps, making the wind powerless. For insect-borne pollen, on the other hand, stickiness is a most desirable attribute, and we do indeed find that such grains are covered with an "adhesive," a glue formed by the anthers. Actually, it is a thick, sticky oil which is not volatile and does not turn into a hard varnish, but retains its fluidity for quite a while. It was a startling discovery, then, when F. Pohl, in 1920, found that traces of the oil are present in a number of wind-pollinated plants in the group of the Ranales, which we mentioned earlier as a primitive group among the flowering plants (see Chapter 18). In grasses, which are also wind-pollinated, there is even more than a trace, although the oil here is thin and thus not very sticky.

We are forced to conclude that wind-pollination in the higher plants is a *secondary* thing—some plants had to give up their partnership with insects, to return to the old alliance with the wind that had been so successful with their ancestors. In some groups, the transition back to wind-pollination was complete and successful; other groups may have been unable to carry out the about-face and must have disappeared from the surface of the earth. In still others, we can observe the transition going on today. Most of the true maples (*Acer* species), for example, are insect-pollinated. Boxelder (*A. negundo*), however, is wind-pollinated. Tropical oaks and chestnuts are at least in part insect-pollinated; those of the more temperate regions are not, with the exception of the edible chestnut, *Castanea sativa*, mentioned in Chapter 18.

Apart from the lack of glue on the pollen grains, a number of other requirements have to be met in order to make wind-pollination successful. For one thing, the pollen has to be very small and light, otherwise, it would sink to the ground too fast. It is a fact that when we compare a large number of wind-borne and insect-borne pollens, the latter are, on the average, bigger (Fig. 84). We have already mentioned (page 87) that there are some notable exceptions such as the pollen grains of the insect-pollinated *Myosotis silvatica* which measure only 0.004 by 0.006 mm—but then we should remember that size, although very important, is only one factor among many. In most alders, hazels, and junipers we find that the average size of a pollen grain is only about 0.030 mm. In still air, such minute, individual grains will not sink down more than an inch per second. To cover a vertical distance of a yard, 36 seconds or a good ½ minute would thus be required. Supposing that there is a wind blowing with a velocity of 7 yards per second, which is by no means unusual, then the pollen would, in those 36 seconds, cover a horizontal distance of a good 250 yards!

By experiments and calculations it can easily be shown that bigger pollen grains, or pollen grains that tend to clump, are at a terrible disadvantage. When, in spite of this, we sometimes find bigger grains in some of the wind-pollinated plants, we notice that Nature has taken special measures to counteract the unfavorable weight factor. Thus, in pine and spruce, we find pollen with air sacs. In larch, where the pollen grains are at least twice as big as they are in hazel—and that is much too big for comfort—such

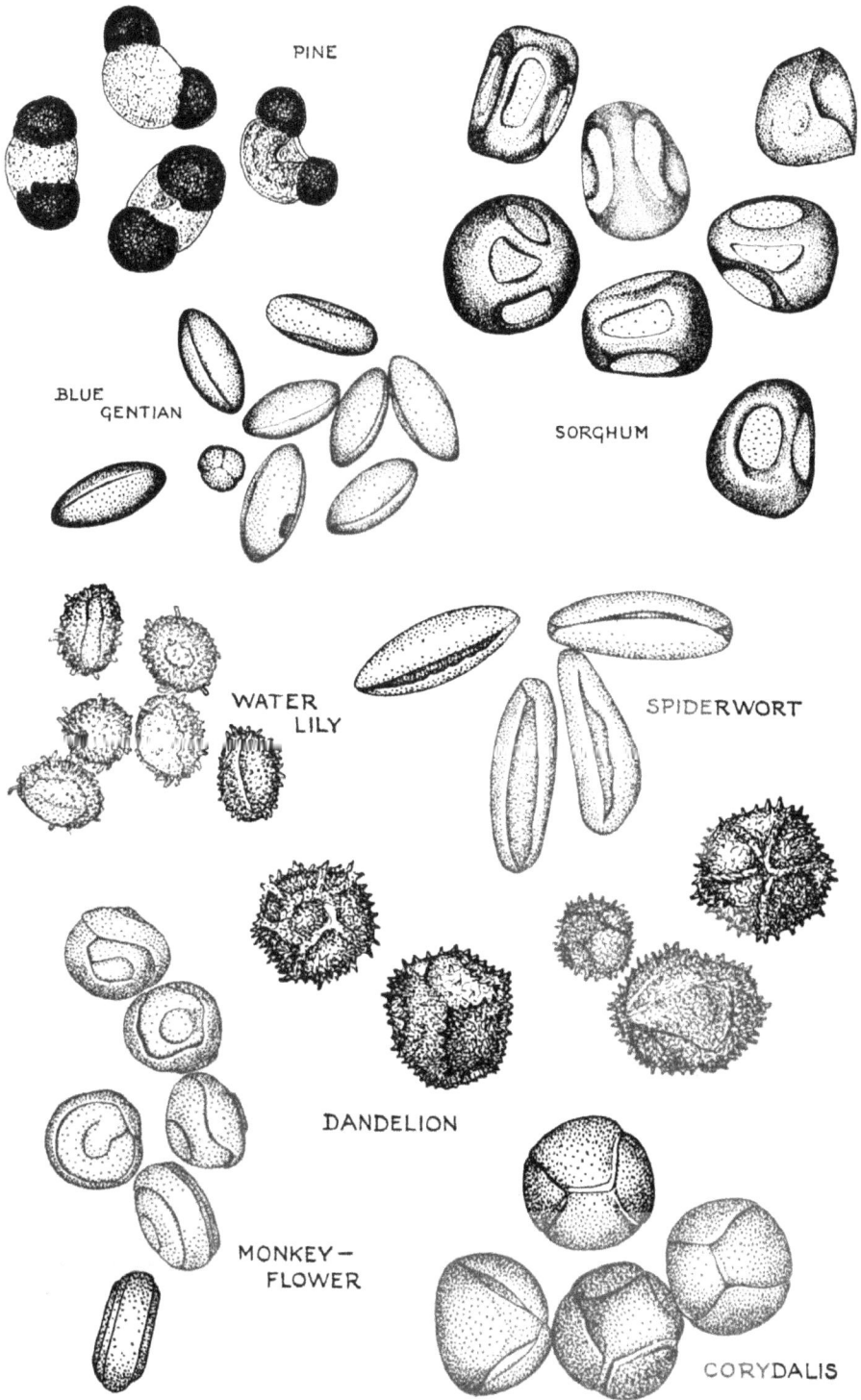

Figure 84. Various types of pollen grains, highly magnified.

sacs are lacking. Here, however, we find that the grains, once they are exposed to air, lose water very rapidly and become more or less watch-glass-shaped, which is a great help in floating. The water loss is not serious, and will of course be remedied automatically once the grains are on the proper, moist stigmas.

Since the chance for a particular pollen grain to find the right stigma is dwindlingly small, the number of grains that the plant produces is simply fantastic. Knoll has calculated that a male hazel catkin yields a good 2 million pollen grains, and this is a conservative estimate. For a hazel bush of average size, with 300 such male catkins, this would amount to a total production of at least 600 million pollen grains. Small wonder that some people get hay fever from this! Also, in connection with the low pollination chance, it would not make sense for a hazel shrub to produce too many female flowers; what counts is the *ratio* between the male and female sex cells that are produced. By actual count, Knoll has found 214 female catkins or 1,284 female flowers on a shrub, which would be equivalent to 2,568 egg cells (two per flower). If nothing went wrong, then, $600,000,000 \div 2,568$, or 233,644 pollen tubes, would be available for each egg cell. That is, over a quarter of a million.

To cut down on the wastefulness as much as possible, most wind-pollinated plants release their pollen only when conditions are favorable, and even then they do not always do it in just one shot. Alder, hazel, and many other wind-pollinated hardwoods produce their flowers very early in the spring, before there is too much foliage that would intercept the pollen. Rain would not only bring the pollen dust down but would also spoil it, and for that reason it should only be liberated when the weather is good and the air dry. Here we notice something like automatic control; the cracks in the anther that make escape of the pollen possible are the natural result of a drying-out process, so *of course* the grains will only escape in good weather! In hazel, a gaping hole will be formed over the entire length of the anther, whose volume is thereby reduced to less than a third of what it was. The pollen has no place to go and must come out. If there is no wind, it does not fall to the ground immediately but simply remains on the neighboring parts of the catkin until the first breeze comes. We find this economy principle not only in plants with hanging catkins but also in those with stiff male inflorescences or flowers, such as cedar, fir, pine, and yew.

Stinging nettles and some of their relatives, such as the Japanese paper mulberry (*Broussonetia papyrifera*) and the little *Parietaria* that is sometimes grown in our greenhouses, have an explosive mechanism to shoot the pollen up into the air. *Parietaria* is proterogynous, that is, the pistil reaches maturity earlier than the stamens. It is only after the star-shaped stigma has sloughed off that the mechanism comes into action. The filaments that were originally folded up in the flower suddenly pop out with considerable force. Each time this happens, a little cloud of pollen is released from the now-ripe anthers.

Grasses have anthers that operate on what we might call the hourglass principle. They dangle from the flower, suspended by slender filaments, and open in good weather because a hole is formed at their lower end. Some pollen emerges from this hole, but in the absence of wind it simply accumulates on a small, protruding horizontal ledge at the base of the anther. When the wind finally comes, the pollen is blown away, and the anther is swung back and forth, with the result that some more pollen will emerge from the "hourglass"—but only a *little* more!

The stigmas in wind-pollinated plants are usually larger, sometimes filamentous, as in box-elder; sometimes lobed, as in walnut; or finely feathered, as we see in the grasses. In hazel, we notice that several filaments from the female flowers that make up one inflorescence have been combined to give a brush-shaped structure.

Calyx and corolla in wind-pollinated plants lack color, that is to say, they are green like the foliage. As a rule, they are not well developed and this can go so far that the flowers are entirely naked. Nectar and fragrance are practically always absent. It is perfectly clear that these flowers are not interested in insects.

## POLLINATION BY WATER

In general, water has a very bad effect on pollen; therefore, it is not amazing that few plants make use of it in pollination. Or perhaps I should say that few plants know how to cope with the problems that arise from the presence of that pernicious substance, water.

One of the most extreme cases is probably that of mermaid weed, *Ceratophyllum,* a plant well known to aquarium lovers. Here, the whole process of pollination takes place under water. The pollen cannot honestly be called a powder, for it comes in the form of tiny, sausage-shaped balloons. About a thousand of these can form quite a cloud in the water, around the flowers which are swayed back and forth through it with their stigmas stretched out. The balloons that are caught on the stigmas will produce pollen tubes that grow down through the pistil, just as we have seen in land plants.

Even more wonderful is the chain of events in that other beloved aquarium plant, *Vallisneria spiralis* or ribbon-weed, a native of northern Italy where it is found in ponds and lakes. Here we have separate male and female plants, just as in the case of date palms. If you check the vallisnerias in your aquarium regularly, you may some day find one which has begun to form two or three buds at its base, in the axil of a leaf. As a rule, one of these buds will turn out to be a leaf bud, which starts to grow out horizontally. The others, however, are regular flower buds which grow up in a more or less vertical fashion. If your plant happens to be a male, the buds will open while still under water. On a short, common stem we then find a bunch of tiny, round pollen flowers, which for the time being behave like real clams and remain tightly closed. If the plant is a female, each bud contains only one little flower, enveloped in some sort of thin membrane. This flower also remains closed, at least for the time being. The long stem on which it is placed continues to grow to the surface of the water, describing a spiral as it does so. On the surface, the female flower gradually opens, showing a number of tiny white "petals" and three pistils that are split in two. The pistils are rather big for such a small flower.

By this time, the tiny spheres which are the male flowers are released one by one; they soar to the surface, on which they will float. It is here that they will open, at long last, and it now turns out that their three "petals" are canoe-shaped and act as pontoons, keeping the two stamens in an erect position above the water. Even when a gentle breeze moves the tiny pollen boats across the surface of the water, there is little danger that the clump of big, sticky pollen grains in the anther will be wetted. If everything proceeds according to Nature's plans, the pollen boats will eventually bump into the female flowers, anchored to their mother plant on the bottom by their long twisted stalks. The jolt of the collision will cause the clump of pollen to be catapulted over to the stigma of the female flower, thus effecting pollination. After this, the female flower will be "pulled in" again, because the flower stalk now changes from a corkscrew to something like a real coil spring. Eventually, the young fruit formed from the flower will be so snugly close to the bottom that it can ripen safely.

In the case of the eel grasses such as *Zostera marina,* found in the shallow sea water along our coasts, the pollen grains float on the surface of the water, and the stigmas are stretched out along that surface. When you come right down to it, it is clear that the wind takes care of the pollination here, and in the case of ribbon-weed as well. So we see that Nature has in an admirable way conquered the problems that can arise when higher plants have to live in an aquatic environment.

# An Epilogue and A Creed

At the end of this book it behooves us to look back and consider our subject in the light of what we have learned. Is the study of pollination really more than what the Germans call a *Spielerei*, a fooling-around that serves no good purpose whatsoever?

Answering such a question is always a little difficult, for it is only too easy to become overserious, with all the ludicrous consequences thereof. It is tempting in this connection to speak of that good old Pennsylvania professor who founded a society for the protection of bumblebees. With just a little more than gentle persuasion, he brought his students so far that they would stand up solemnly in class, raise their right hands, and swear that they would protect and honor bumblebees for the rest of their lives. I am afraid that, for all his good intentions, this professor overshot the mark a little bit. He took himself too seriously.

Yet it does not make sense to deny that there is at least a core of seriousness in our subject. Although I have thus far refrained from saying so in so many words, the story of pollination is a glowing testimonial to the theory of evolution. In the dim, gray past, certain flowers and animals began to march through time together. They became more and more dependent upon each other, and finally some partnerships evolved that today fill us with marvel and admiration.

It is too bad that Sprengel did not know about evolution. Familiarity with the idea would perhaps have helped him to get over his disappointment when he observed the burglarizing bumblebees. Does not their behavior show very definitely that this is not the best of all possible worlds? Conversely—and there is a most gentle irony in this thought—after a period of rather crass materialism, we are brought to an attitude closely akin to Sprengel's by our very belief in evolution and especially by the idea that it applies to man. It is a religious belief in the sense that it imposes upon us moral obligations and restrictions not different from those recognized by people who are "religious" in a more conventional way. True, we have no wish to deny our past, but neither do we wish to deny or jeopardize our future. If we feel some modest pride in stating that man is more than monkey, we must live up to our responsibilities—especially at this crucial moment in the history of mankind with the lurking danger of atomic war and radioactive contamination.

Let the unselfish study of Nature be a help to us in retaining our sanity and in preventing us from doing the evil things that would break our particular line of evolution forever. In short, let us admire Creation; let us not destroy it. If this book has contributed toward that goal, be the contribution ever so little, it will have fulfilled its purpose.

# A Selected Bibliography

BAKER, H. G., and B. J. HARRIS. 1957. The pollination of *Parkia* by bats and its attendant evolutionary problems. *Evolution* 11: 449–460.

BOHART, G. E. 1952. Pollination by native insects. In A. Stefferud (ed.). *Insects: The Yearbook of Agriculture, 1952.* Government Printing Office, Washington, D.C. Pp. 107–121.

BUTLER, C. G. 1949. *Honeybee, an Introduction to Her Sense-psychology and Behavior.* Oxford University Press, New York.

CARPENTER, G. H. 1928. *The Biology of Insects.* Sidgwick & Jackson, Ltd., London.

CARPENTER, G. D. H., and E B. FORD. 1933. *Mimicry.* Methuen & Co. Ltd., London.

CARTHY, J. D. 1956. *Animal Navigation.* Charles Scribner's Sons, New York.

CLEMENTS, F. E., and F. L. LONG. 1923. Experimental pollination. An outline of the ecology of flowers and insects. *Carnegie Inst. of Washington Publ. 336.*

COMSTOCK, J. H. 1917. *How to Know the Butterflies.* D. Appleton & Co., New York.

COTT, H. 1957. *Adaptive Coloration in Animals.* Methuen & Co. Ltd., London.

COWAN, F. 1865. *Curious Facts in the History of Insects.* J. B. Lippincott Co., Philadelphia.

CURRAN, C. H. 1951. *Insects in Your Life.* Sheridan House, Inc., New York.

CURTIS, G. D. 1948. *Bee's Ways.* Houghton Mifflin Co., Boston.

DARWIN, C. 1877. *The Effects of Cross and Self Fertilisation in the Vegetable Kingdom.* D. Appleton & Co., New York.

———. 1877. *On the Various Contrivances by which Orchids are Fertilized by Insects.* D. Appleton & Co., New York.

EDGELL, G. H. 1949. *The Bee Hunter.* Harvard University Press, Cambridge.

FABRE, J. H. C. 1914. *The Mason-Bees.* Dodd, Mead & Co., New York.

———. 1915. *Bramble-bees and Others.* Dodd, Mead & Co., New York.

FORD, E. B. 1955. *Moths.* The Macmillan Co., New York.

FRISCH, K. VON. 1955. *The Dancing Bees.* Harcourt, Brace & Co., Inc., New York.

———. 1956. *Bees: Their Vision, Chemical Senses, and Language.* Cornell University Press, Ithaca, N.Y.

GRANT, V. 1949. Arthur Dobbs (1750) and the discovery of the pollination of flowers by insects. *Torreya (Bull. Torrey Bot. Club)* 76: 217–219.

———. 1950. The flower constancy of bees. *Bot. Rev.* 16: 379–380.

———. 1951. The fertilization of flowers. *Scientific American* 184: 52–56.

GREENEWALT, C. H. 1960. *Hummingbirds.* Doubleday & Co., Inc., New York.

———. 1960. The hummingbirds. *National Geographic Magazine* 118: 658–678.

HERTZ, M. 1939. New experiments on colour vision in bees. *J. Exp. Biol.* 16: 1–8.

HOLLAND, W. J. 1931. *The Butterfly Book.* Doubleday & Co., Inc., New York.

HOWARD, L. O. 1937. *The Insect Book.* Doubleday, Doran & Co., Inc., New York.

HUBER, F. 1808. *New Observations on the Natural History of Bees.* John Anderson, Edinburgh.

IMMS, A. D. 1931. *Social Behaviour in Insects.* Methuen and Co. Ltd., London.

JAMES, W. O., and A. R. CLAPHAM. 1935. *The Biology of Flowers.* Oxford University Press, New York.

KERNER, A. 1895. *The Natural History of Plants, Their Forms, Growth, Reproduction, and Distribution.* Trans. by F. W. Oliver. Henry Holt and Co., New York.

KERR, W. E. 1960. Evolution of communication in bees and its role in speciation. *Evolution* 14: 386–387.

KHALIFMAN, I. 1955. *Bees: a Book on the Biology of the Bee-colony and the Achievements of Bee-science.* Lawrence & Wishert, London.

KLOTS, A. B. 1951. *Field Guide to the Butterflies.* Houghton Mifflin Co., Boston.

———. 1958. *World of Butterflies and Moths.* McGraw-Hill Book Co., Inc., New York.

KNUTH, P. 1906–1909. *Handbook of Flower Pollination.* Five vols. Trans. by R. A. Davis. Oxford University Press, New York.

LOVELL, J. H. 1918. *The Flower and the Bee: Plant Life and Pollination.* Charles Scribner's Sons, New York.

LUBBOCK, J. 1913. *Ants, Bees and Wasps.* D. Appleton & Co., New York.

LUTZ, F. E. 1935. *Field Book of Insects.* G. P. Putnam's Sons, New York.

———. 1941. *A Lot of Insects: Entomology in a Suburban Garden.* G. P. Putnam's Sons, New York.

MACY, R. W., and H. W. SHEPARD. 1941. *Butterflies.* University of Minnesota Press, Minneapolis.

MAETERLINCK, M. 1954. *The Life of the Bee.* New American Library, New York.

MANNING, A. 1956. The effect of honeyguides. *Behaviour* 9: 114–140.

———. 1956. Some aspects of the foraging behaviour of bumblebees. *Behaviour* 9: 164–203.

MÜLLER, H. 1876. On the relation between flowers and insects. *Nature* 15: 178–180.

———. 1883. *The Fertilisation of Flowers.* Trans. and ed. by D'Arcy W. Thompson. Macmillan and Co., London.

NEEDHAM, J. G. 1943. *Introducing Insects.* The Ronald Press Co., New York.

NIXON, G. 1955. *The World of Bees.* Philosophical Publishing Co., Quakertown, Pa.

NOAILLES, R. H., and J. M. GUILCHER. 1954. *The Hidden Life of Flowers.* Philosophical Publishing Co., Quakertown, Pa.

PEARSON, O. 1953. The metabolism of hummingbirds. *Scientific American* 188: 69–72.

PESSON, P. 1959. *The World of Insects.* Trans. by R. B. Freeman. McGraw-Hill Book Co., Inc., New York.

PHILLIPS, E. F. 1928. *Beekeeping.* The Macmillan Co., New York.

PIERCE, G. W. 1948. *The Songs of Insects.* Harvard University Press, Cambridge.

PLATH, O. E. 1934. *Bumblebees and Their Ways.* The Macmillan Co., New York.

POULTON, E. B. 1890. *The Colours of Animals.* D. Appleton & Co., New York.

RANSOME, H. M. 1937. *The Sacred Bee in Ancient Times and Folklore.* Houghton Mifflin Co., Boston.

RÉAUMUR, R. A. F. DE. 1744. *The Natural History of Bees.* J. P. Knapton, London.

RIBBANDS, C. R. 1957. *The Behavior and Social Life of Honeybees.* Dover Publications, Inc., New York.

ROOT, A. I. 1950. *The ABC and XYZ of Bee Culture.* Rev. by E. R. Root *et al.* A. I. Root Co., Medina, Ohio.

ROSS, E. S. 1953. *Insects Close Up.* University of California Press, Berkeley.

SHARP, D. L. 1925. *The Spirit of the Hive.* Harper & Bros., New York.

SLADEN, F. W. L. 1912. *The Bumble-bee.* Macmillan & Co., London.

SNODGRASS, R. E. 1925. *The Anatomy and Physiology of the Honeybee.* McGraw-Hill Book Co., Inc., New York.

———. 1956. *The Anatomy of the Honeybee.* Cornell University Press, Ithaca, N.Y.

STEP, E. (ed.). n.d. *Marvels of Insect Life.* Hutchinson and Co., London.

TEALE, E. W. 1954. The journeying butterflies. *Audubon Magazine* 56: 206–211, 230–231.

———. 1960. *A Book About Bees.* Dodd, Mead & Co., New York.

THORPE, W. H. 1949. Orientation and methods of communication of the honeybee and its sensitivity to the polarization of light *Nature* 164: 11–14.

TINBERGEN, N. 1953. *Social Behavior in Animals.* Methuen & Co., Ltd., London.

———. 1958. *Curious Naturalists.* Country Life, Ltd., London.

URQUHART, F. A. 1960. *The Monarch Butterfly.* University of Toronto Press, Toronto.

WHEELER, W. M. 1923. *Social Life Among the Insects.* Harcourt, Brace & Co., Inc., New York.

———. 1928. *The Social Insects, Their Origin and Evolution.* Harcourt, Brace & Co., Inc., New York.

WILLIAMS, C. B. 1958. *Insect Migration.* The Macmillan Co., New York.

# Index

Winter aconite, *Eranthis hiemalis*, 78
*Wisteria*, 110
Wolff, T., 163
Wollschweber, *Bombylius fuliginosus*, 22
Workers
  in bumblebees, 105
  in honeybees, 119, 127

*Xylocopa; see also* Carpenter bees
  and biduri flowers, 143
  fooled by *Canavalia* flowers, 117
  occurrence, 112, 113
  as pollinator of Scotch broom, 51
*Xylocopa virginica*, 112

Yeasts (in nectar), 80

Yellowjacket (*Vespa* species); *see also*
    Wasps
  color vision, 22, 27
  as models for Syrphid flies, 150, 153–55
Yew, 214
*Yucca*, 190–91; *see also* Spanish dagger
    plant
Yucca moth, *Pronuba yuccasella*, 190, 191

Zebra caterpillars, 154; *see also* Cinnabar
    moth
*Zinnia*, facing 30 (Fig. 9), 31
*Zostera marina*, 217; *see also* Eel grass
*Zosteropidae*, 96; *see also* Spectacle-birds
Zucchini, dependence on cross-pollination,
    7

www.ingramcontent.com/pod-product-compliance
Lightning Source LLC
Chambersburg PA
CBHW051716020426

42333CB00014B/1003